Reality through the Looking-Glass

Christopher Clarke is Professor of Applied
Mathematics and Dean of Faculty at the University of Southampton. His research is
divided between large-scale structure of the
universe and the physics of the brain and
consciousness. As a Christian, he is closely
associated with the creation spirituality
movement.

C J S Clarke

Reality through the Looking-Glass

Science and awareness in the postmodern world

Floris Books

First published in 1996 by Floris Books

British Library CIP Data available

ISBN 0-86315-216-3

Printed in Great Britain
by Cromwell Press, Wilts.

Contents

'What,' it will be questioned, 'when the sun rises, do you not see a round disc of fire somewhat like a guinea?' 'O no, no, I see an innumerable company of the heavenly host crying, "Holy, Holy, Holy is the Lord God Almighty".'

William Blake
A Vision of the Last Judgment (1810)

I have ... drawn up my chairs to my two tables. ... One of them has been familiar to me from earliest years ... it is coloured; above all it is substantial. ... Table No. 2 is my scientific table. ... My scientific table is mostly emptiness. Sparsely scattered in that emptiness are numerous electric charges rushing about with great speed.

Arthur Eddington
The Nature of the Physical World (1928)

Acknowledgments

Out of the many people who have influenced this book, a special thanks must go to Isabel Clarke whose ideas about the *nagual* play a central role, and to Jacques Vroemen whose gentle approach to shamanism influenced many finishing touches. I must also thank all the members of my Southampton University course on 'Reality' whose patience and enthusiasm provided the impetus for starting this book.

I gratefully acknowledge the permission of Threshold Books, Putney, Vermont, to reproduce 'Be Melting Snow,' from *The Open Secret: versions of Rumi,* translated by John Moyne and Coleman Barks, 1984.

Introduction

Most of us think we have a reasonable grasp on reality. We have sorted out that Father Christmas/Santa Claus is unreal, along with unicorns and fairies. We suspect that those who think they are being manipulated by death-rays beamed from Mars have lost their sense of reality, and we assume that it is the business of science to find out, and tell us, what reality consists of. We assume that when we are awake, sober and healthy then we are fully in touch with reality; but with dreams, drunkenness, drugs or delirium, we enter unreality.

This view is regarded as modern. Those who think this way regard older ideas as superstition. They contrast modern thinking with thinking in earlier times, when people were confused about such matters, believing that dreams could tell one about reality, or that myths were as reliable as science. It is often said that the transition between superstition and modern thought took place in the eighteenth century, in a process often called the Enlightenment, when society woke up and saw the truth about reality, sweeping away the darkness that went before. We moderns now live with a sure idea of the way things are, which is the only foundation for building the future.

Today, however, this modern view is under attack from all sides. Many of those mounting the attack call themselves postmodernists, anxious to stress that by attacking the modern view they are not simply going back to earlier ways of thought. They argue that what counts as reality varies from one culture to another, and that we have to acknowledge that the world now consists of many different cultures with many conflicting views of reality which have to be recognized. Quite suddenly the old ideas seem to many to have outgrown their usefulness, and we are faced with the need to revise them, to set them in a wider context.

A more subtle attack comes from parts of science, which now suggest to some that scientists are not simply discovering a reality that already exists, but are rather creating some particular theoretical picture of the world as a result of their activities. These ideas amount to a revolution in scientific thought as momentous as the seventeenth century revolution that established the modern view.

As a result, there is a growing need to replace the 'modern' set of assumptions, a need to alter the foundations of our thinking. Every sphere of life is affected by this: it alters not only the way we do science, but the way we educate our children, the way we treat each other, the way we treat the world. If our accepted assumptions about reality collapse, what will replace them as a basis for life?

In the chapters that follow, I will describe these attacks on the old notions of reality and the way in which a new view of the world can emerge. In Chapter 1, I relate the modern sense of reality to science, and describe the postmodern argument, that science depends on society, so that when society changes, so does science, and so does the sense of reality. This way of thinking forms the foundation for everything that follows.

I also touch in this chapter on a way of thinking that many writers have linked with postmodernism, namely feminism. The feminist critique will play a key role in understanding the evolution of modern science — as has been stressed by E. Fox Keller (1985). At this stage I only note some of its features, including its introduction of a moral dimension into the discussion, in which morality is understood in terms of responsibility and relationship, not in terms of rules and prohibitions.

Chapter 2 starts to examine, from this postmodern foundation, the legacy of scientific realism that we have inherited from the past. In keeping with the principle that science is moulded by society, I look for the roots of our scientific world-view in the changes in society that have taken place over the centuries. The central historical event that I need to explain is the decision, taken in the seventeenth century, to base science, and the scientific conception of reality, on the theory of atoms. Remarkably, this only becomes comprehensible from a very broad perspective, which sees this event as the culmination of the progressive consolidation of patriarchal power, which overcame the previous matricentral society of Europe in the middle Bronze Age. Even more surprisingly, this enlarged perspective reveals the moral dimension that I have noted is central to the feminist critique. What might have been a philosophical discussion suddenly starts to assume human significance. From this perspective, I describe the way in which thinkers in the seventeenth century set up the scientific view of reality that we have today, and how they made a complete separation between the ordinary world of our senses and the theoretical world of scientific reality.

The next chapters take up the criticism of the seventeenth century world-view that has emerged from within science itself, based not on considerations of the nature of society, but on the internal consistency of the theory of atomism. Chapter 3 begins this process by examining the debate about the nature of space and time that surrounded the adoption of atomism, and which has continued ever since. The whole of modern science rests on ideas of space and time, and yet they have been surrounded from the start by great uncertainty.

Then in Chapter 4 we reach modern physics, where further problems are raised about the conventional atomistic world view by the phenomena of quantum theory. The significance of this chapter is twofold. First, it further undermines the certainty of atomism. Second, it starts to develop an alternative way of looking at reality based on quantum theory, a way that focuses on relationship rather than on separateness, and so starts to alter that moulding of science which took place as a result of the dominance of patriarchy.

The central point of our investigation is reached in Chapter 5, on consciousness. This is a theme that modern science has tried to ignore, but which is now pressing back irresistibly. It raises the most fundamental questions about the nature of reality and the nature of the person.

Consciousness is a many faceted, and much misunderstood, concept. The very idea is derived to a large extent from a traditional scientific world view, which makes a separation between the external world and the 'me' that experiences it. Connecting the world with me is this thing called consciousness, the postulated faculty within a human being for registering both the external world and the internal world of thoughts and feelings into a single composite awareness. The scientist approaches the question by asking for the mechanism whereby the brain carries out this conversion from signals, coming from the world and the body, into our awareness.

The philosopher or psychologist, on the other hand, may turn the problem round: starting with my awareness, which is the world, the reality, *for me,* we ask how this single awareness comes to be split up into a separate 'me' and 'the world.' In such an analysis there may not be a place for consciousness as a distinct human faculty.

We have to understand consciousness if we are to understand the relation, or lack of it, between scientific reality and the 'common sense reality' of our ordinarily experienced world. Some of the things in this ordinary world seem to fit well with the scientific approach: the way in which external objects are laid out in space; the laws that objects seem to obey, and so on. But some things seem to be in a completely different world from that of science. Not only are our deepest emotions in this category, but even certain aspects of the external world, such as the particular subjective experience of a certain colour. Science can explain what sort of vibrations of light correspond to this experience, but to explain the experience itself, and what it has to do with vibrations of light, science is powerless. These primitive experiences, called by philosophers *qualia,* are a key issue in understanding consciousness, the reality of the world, and the nature of human experience, which I explore in detail in 5.2.

The main philosophical conundrum about qualia is the question of whether my qualia are the same as your qualia. If we are both looking at a red curtain, is the experience that I describe as 'red' the same experience as the one that you describe as 'red'? The problem is unanswerable on all conventional approaches to science and philosophy. On the approach being developed in this book, however, qualia fall naturally into place and the question can receive an answer.

With the consideration of consciousness, a new picture of reality emerges in Chapter 6, 'A Quantum World?' The essence of this approach is that we cannot start with the picture of reality that science has constructed and then try to recover consciousness, qualia and the rest of experience. Rather, we must start with experience, with the world as it is first given to us, and recognize that many different strands of analysis can then be drawn out of this experience. Indeed, the analysis is inseparable from the experience; there is no 'raw experience' that we can start from. One analysis is that of science, in its conventional form, which provides a vital insight into certain sorts of experience, but leaves much of the world untouched. This provisional nature of science has been revealed through the preceding chapters. Other approaches will be considered later on. Central to all analyses is the distinction between 'me' and 'not-me,' subjective and objective. The distinction is, however, not absolute. There are no static objects, called the Self and the World. Instead there is a web of shifting relationships in which, by a strange inversion of usual logic, the relationship comes before the things related.

In this new picture of the world, qualia are not things that mysteriously emerge within individual private minds, but are to some extent real parts of the world. In such a picture it is now possible to make sense of saying that two qualia, such as my experience of red and your experience of red, are the same. More precisely, we will have reached a view in which there are not two qualia, one in me and one in you, but one quale that we both share. Then our sensations are the same because any thing is the same as itself. This view depends on the idea that (with significant qualifications) qualia are not secondary qualities, added in the brain to an object that is 'really' colourless, but that an important component of the quale resides in external objects themselves, and so is primary. In a sense, it is a return to the naïve view of reality that was overturned by Descartes; but this will only be possible through a picture that takes full account of all we have learned about the human brain, and what it contributes to our view of the world, in the intervening period since Descartes.

I describe this view of the world as a quantum view, not just because it is using ideas from quantum theory, but because there is a way of generalizing quantum theory itself so that it provides a language that is equally capable of describing fundamental particles and the shifting patterns of human awareness. This 'broad quantum theory' is described in Chapter 6.

So far I have described two attacks on the conventional scientific view of reality: the attack of postmodernism, from the side of sociology and history, and the attack of quantum theory, from within science itself, and I have indicated through considering consciousness how quantum theory opens up a new alternative to the Newtonian view in which the world is a much richer place than is the case in conventional science.

Here, however, I must face up to a central criticism that has been levelled at many writers who link quantum theory with consciousness and/or postmodernism. It is argued that these are quite different things, and that it is misleading and dishonest to confuse them. Quantum theory, it is argued, is a very specific physical theory, built within the modern scientific understanding of the universe, aimed at explaining the behaviour of atomic particles; consciousness belongs to a branch of psychology, and so is equally a part of modern science, but one having little to do with quantum theory; and postmodernism is a critique of the foundations of modern science, and so at a quite different level from either consciousness or quantum theory. On this view postmodernism cannot possibly draw support from quantum theory because postmodernism itself undermines the truth of quantum theory, as it does the truth of all modern science and, indeed, the very notion of truth itself.

My answer lies in the quite radical way in which I am viewing all these components in my argument. I have already noted how quantum theory, as will emerge in Chapter 6 can be understood not as a particular physical theory but as a new sort of language for describing the world. It is a language that is capable of talking about the provisional, shifting, context-dependent world that is opened up by the postmodern analysis. From the point of view of broad quantum theory, the quantum language is the one that is natural for dealing with the world as it is, and classical physics results from imposing stringent metaphysical restrictions on this natural language.

Postmodernism, in its turn, is more than just a way of studying the philosophy of science. It shows that the social and historical context determines, not just the formal theories of remote research laboratories, but the whole pattern of thinking of the society in question. The historico-social context influences our theories, our values, our concepts and, through these, our very perceptions. Consciousness is socially conditioned, and with it is conditioned our whole world and our concept of reality.

Consciousness, finally, is far more than a branch of psychology. It *is* our world; it is reality. Any account of the world that fails on consciousness fails totally. Yet we have seen that the most basic parts of consciousness, the qualia of our awareness, depend completely on quantum theory for their explanation. Far from residing in separate and incompatible areas of human discourse, the three subjects of consciousness, postmodernism and quantum theory are so intertwined that it is impossible to speak of one without the others.

So we reach a picture of the world in which the nature of reality is profoundly altered by the juxtaposition of these three angles of approach. Yet we still have not answered the most fundamental question of all: what is reality? The Newtonian picture attempted to answer this once and for all by postulating a purely abstract scientific world view based on the doctrine of atomism. Now that this has failed, are we left with no alternative but to abandon any mention of 'reality'; no alternative to a complete relativism in which each person and each society creates their own equally valid personal 'world'? Is it the case that, without qualification, anything goes?

For me, the answer to this must be, no. The scientific enterprise, while it is very far from a passive recording of a reality that is entirely independent of ourselves, is certainly not a free composition of our imagination. Our interaction with a world that is in some respect distinct from ourselves is absolutely essential for science and for our existence as mature human beings. We need, therefore, a way of operating in this situation where it is essential to recognize a dimension of existence that is distinct from ourselves, and yet where everything that we say about this dimension is conditioned through and through by our own creativity.

This dimension of existence that is distinct from ourselves is, however, not separate from us; it is beyond every description of the world and yet is the foundation and source of every description. Even to say 'it' of this dimension is to beg the whole question, since our language is so structured that every 'it' is a self-contained and objectively existent entity.

It is no accident that, in trying to answer the question: what is reality? I have been led into language reminiscent of religious mysticism. For I would contend that this is precisely where the answer is to be found. Not as a court of last resort or desperation, but as the natural framework within which human experience can shed light on the fluid situation presented by the breakdown of the old realities. The discipline of mysticism provides a way forward that recognizes the inherent limitations of every language — even the artificial language of mathematics that can achieve so much by freeing itself from everyday reference points.

As a result, the final chapters are concerned with an extension of the discussion of reality into the territory of religion. In 'Alternative Realities' we first look at religions that are usually called 'primitive,' but which, in the light of the historical picture that I have been building up, can be seen as reaching back to a time before the patriarchal moulding of our world-view. These earlier conceptions of the world raise the possibility of viewing the world in quite different ways, and hence the possibility of alternative realities. I give a framework, due to the psychologist Mahrer, for viewing these realities, and from Castaneda I draw some key notions about the status of our provisional construction of reality in relation to whatever lies outside its compass.

The last chapter then tries to point towards this unknown realm that is beyond our construction of reality and yet is its precondition, by looking specifically at religious reality. The term might be misleading: a 'religion' is a particular social structure, and we have already seen that each such structure constructs its own, purely relative, reality. In this sense, it might be better to speak of 'mystical reality,' pointing to that level of experience within the religious traditions at which direct experience goes beyond the particular formulations that characterize the religion. But I continue to speak of religious reality, in order to emphasize that there is no such thing as 'generic mysticism,' whatever some have said to the contrary. What I referred to above as the discipline of mysticism is its commitment to a path that is situated among the actualities of particular religious traditions and languages. It is through these languages, with their very specific cultural and historical roots, that mysticism points to what is beyond language.

Books, like societies, have historical origins. This book stemmed from a one-term adult education course entitled 'Reality,' where a group from all manner of backgrounds challenged and shaped my emerging ideas. It stems also from my own commitment to the Christian spiritual path (currently leading me through the dark vale of the Church of England). The influence of this path, particularly in the form of creation spirituality, can be seen at many points. Indeed, not only the final chapter, but the whole book could be seen as an expression of this religion, which I see not as a collection of stories about God, but as humanity's feeble response to a world in the process of becoming, a world that strips us of every certainty except for the presence at its heart of an unbounded compassion.

Chapter 1
Reality and Beyond

1. The concept of reality

The word 'really' is constantly on our lips. It stands for the true, the reliable, the authentic. The quest for reality is the quest for a final assurance of absolute truth on which we can base our lives and our society. What is this concept, on which we stake so much? How do we use the word, and what are we implying?

We often compare *reality* with *appearance*. When a conjuror apparently passes a sword through his beautiful assistant we ask, 'what really happened?' When a shimmering lake appears in the desert we may suspect that it is 'really' a mirage. Here we are contrasting two realms: the realm of illusion, of appearance, which we are in if we follow our senses uncritically; and another realm reached by probing behind our senses, questioning whether what we see fits in with the way our intellect suggests the universe might be. In this sense, reality is an intellectual matter, something that arises from applying reasoning to what we at first perceive.

In other circumstances the comparison is between reality and imagination. Here the focus is not so much on what lies behind our experience, but whether or not something has been experienced at all. When we inquire whether a book we have read is a novel or an auto-biography, we are asking whether or not it is 'real,' meaning, whether the author had the experiences described, or whether they were invented to make a good story. If the experiences are totally different from anything we could expect to experience, then we might suspect it to be imaginative fiction, saying that 'in reality' such things do not occur. On the other hand, when we assert that a novel is very 'realistic,' we mean that there is no reason why the experiences should not have been had by the author.

In all these cases, we are matching what we see or what we read with some definite idea of the way the world is. It is this basic *what the world is like* that we call reality. The world, our world, does not on the whole show instances of people happily smiling as swords are passed through them, and so we suspect the conjuror of giving us an illusion, an experience not corresponding to reality. In our world we meet many selfish people who harm others for their own ends, and so a novel full of people who behaved with complete altruism might be described as 'unrealistic.'

Thus 'the world,' with which we compare in order to decide whether something is real or not, comprises social experience as well as physical experience. Social experience, however, clearly varies from one society to another. A novel that seems unrealistic from the point of view of European society could be completely realistic if regarded as set in an African society, or vice versa. Many societies, for example, are dominated by the fear that one is in constant danger of being bewitched through the power of the evil eye (Taussig 1986, p.172), whereas in Western Europe such thoughts are restricted to a very small minority who tend to be referred for psychiatric treatment.

A key assumption of modernism, however, is the idea that, whereas the behaviour of people — the social world — may vary from one culture to another, there are some things that are independent of culture. That stones fall towards the earth when released is a fact on which every culture will agree, even if they disagree on its explanation. Facts like this, it might be thought, constitute an absolute reality, a realm of things that are not merely real for me or real for you, but 'really real'!

This is the realm of science. There is much that can be said about what is and what is not scientific; but probably all scientists would want to maintain that the subject matter of science should consist of facts that are real in the sense of being agreed on by all cultures. Moreover, scientists would regard it as their job to determine what these facts are. So we see that the concept of reality starts to get bound up with the scientific.

We are now already entering disputed territory, however. For what the scientist accepts as a fact could be different from what all social cultures accept as a fact. As we shall see in Chapter 4, most physicists would regard it as a real fact that matter is made up of objects that sometimes behave like billiard balls and sometimes like waves on the sea. For the vast majority of people, of all cultures, this is not a recognized fact and would be regarded as highly improbable. Thus the scientist would not allow reality to be determined by a democratic vote, even if there was an overwhelming majority. In a case such as this, the scientist would claim that the majority was simply wrong. So we seem to have some facts, like 'stones fall downwards' where reality is a matter of general agreement, and other, more complicated and theoretical facts, where scientists with specialized knowledge are prepared to pronounce on reality even where there is disagreement.

This whole area of the relation between science and society is one to which I will turn later in this chapter. It brings us to the most distinctive part of the modern view: namely the general acceptance that science is entitled to pronounce on the nature of reality; that, in case of dispute as to what might be illusion and what reality, the scientist possesses the appropriate methods for judging the case. In a criminal trial for rape, the evidence of an expert scientist is brought in to present evidence from genetic analysis as to whether or not a suspect is guilty, and the scientist's view carries almost overwhelming weight. In a civil case of compensation for cancers allegedly induced by a nuclear power station, the expert medical statistician is brought in to give a scientific judgment, which will almost invariably be accepted, unless challenged by another scientist. While the competence of the individual scientist may be questioned, the appropriateness of the appeal to science is almost never questioned. Our society is founded on the view that there is a single solid reality, that truth is a matter of agreement with reality, and that science is a procedure for determining the nature of reality and for judging the truth of factual statements.

We so take this state of affairs for granted, that we forget that it is indeed a distinctively modern view. In a 'trial' among the Australian Aborigines, for example, as among many other indigenous societies, the prime weight would have been given to the pronouncement of an accepted expert, the medicine man (Elkin 1954), who could determine the guilty party by dreaming or similar magical means. We would firmly say that such a procedure was wrong and unjust; that we have access to reality while they do not. How far, however, can we be sure of this? Has modern Western society found the key to absolute truth, whereas all previous societies were totally in error, or is the situation more complex than this?

In the rest of this chapter, I shall describe the nature of the modern view of reality, and how it is linked to science, and then examine some of the criticism that has been levelled against this view. The essence of the criticism, particularly that of Paul Feyerabend (see p.29), will be that science is not a privileged hotline to reality, but that it is just one particular sub-culture in our society; a sub-culture that has developed a particular set of ideas in response to a particular historical development that could have been quite different. This will then lead on to the postmodernist view that, if there is no hotline that can ever tell us what reality is, then it is misleading to talk about reality in the first place.

The first aspect to note about the real world of the moderns is that it is impersonal. Matthew Fox (1988) remarked that the main question that humanity asks is: is the universe a friendly place or not? The modern concept of reality holds that it is neither. It is not governed by friendly spirits who encourage our corn to grow, who bring water into wells to refresh us, who inspire us to beautiful thoughts, who lead away our soul after death. Nor is it in the sway of vengeful demons who wreck our ships in the storm, crumble our cities in earthquakes and devastate our crops in the drought. The 'real' world is largely indifferent to us, going its own mechanical way, its blind and random consequences only labelled as weal and woe in our own eyes. We can neither propitiate nor befriend the real world. It goes its way entirely unaffected by our pleas and cries. The only way we can better our lot is by taking control over the world and manipulating it to get the best we can out of it, shaping it to meet our needs.

We need go back only a few hundred years, or move only a little to a different culture, to see how distinctive this view is. In the majority of cultures, for the majority of history, humanity has thought quite differently, holding that the world was essentially animate, as open to influence by me as is another person. It was either pervaded by spirits, or it was under the control of the divine providence of God in all its details.

The change from the old animistic or theistic view to the modern view was very much the result of the rise of science in the late sixteenth and seventeenth centuries. As I shall trace in detail in a later chapter, science proposed a picture of the universe as a vast machine, built up from tiny atoms in the way a clock is built up from cogs. Driven by this picture, science systematically set out to develop a picture of reality from which every human emotion was excluded.

It is interesting that such a desolate picture is not one that most people can live with, so that many look for ways of escape from the remorseless logic of the machine. Some believe in the remote masculine God who emerged from the scientific revolution of the seventeenth century: placed at a great distance outside this pitiless universe, charged with the task of mitigating its harsher consequences and receiving to Himself those souls chosen to be rescued from the uncaring world of matter. Others, perhaps many more in numbers, read their horoscopes, half in jest and half seriously, hoping for a sign that there is some human meaning in the seemingly mechanical rotations of the stars and planets.

Yet alongside this mechanical, scientific picture there are other conceptions which seem to contradict it. While the 'official' view may be the view of the universe as a machine, in many cases it is the human that is brought forward as reality. When I have a piercing toothache, that is the whole of reality for me at that time and everything else fades into unreality in comparison. If I am in love, the whole universe dances with me, or weeps with my unrequited sorrow. We are happy to live with a two-layered reality, the mechanical alongside the emotional, in which our actions are governed by the human reality, but our words by the mechanical, so that the lover will admit that the sun only *appears* to shine more brightly, and that it 'really' is unaffected by human emotion. Even those suffering from what their fellows would classify as delusion retain such a strong sense of the dominance of the scientific view of reality that they will invent pseudo-science to justify their delusions, attributing their misfortunes to the influence of unseen laser-beams or death-rays. This idea of layers of reality will be examined in detail later. The relation between scientific reality and human reality will take us into the whole area of consciousness, explored in Chapter 5; while Chapter 6 will investigate the possibility that there may be yet other layers of reality available to us.

Our first goal, however, must be to understand how the scientific view of reality has gained such prominence. Why should science have the last word on reality? Has science a privileged position from which it can pronounce on reality, or are the pronouncements of science no better or worse than the pronouncements of many other sections of society? If the latter is the case, then the possibility opens up that the dominant view of reality in our culture is founded on the particular historical accidents that caused science to take the path it has. In particular, as we shall explore in more detail (see p.43), our view of reality may be based on a particular development that occurred at the start of the seventeenth century: the adoption of Greek atomism. For, while this proved to be an immensely important tool in yielding understanding of and power over the universe, there is no guarantee that it encompasses the whole of reality, or even that it says anything about the essential nature of any part of reality.

To answer these questions we need to examine the process of science, and the controversies that have grown up around what science is doing. Starting at the end of the seventeenth century, and continuing with increasing force up to the present day, philosophers of science have argued that science is not in fact the sole repository of a gradually perfecting corpus of absolute truth; but is only the particular product of how each generation chooses to represent the world to itself. It is now appropriate to look at some highlights of these arguments.

2. Revolution and paradigms

A key work in this history of ideas is Thomas Kuhn's *The Structure of Scientific Revolutions* (1962). This is the book that introduced the word 'paradigm' into modern speech, and which did much to start a more critical view of the nature of science.

Kuhn's picture of science (or rather, his picture of any particular science, such as physics or biology) is a picture of periods of what he calls *normal science,* punctuated by *paradigm changes.* The two ideas are inseparably linked. He defines normal science as follows:

> In this essay, 'normal science' means research firmly based upon one or more past scientific achievements, achievements that some particular scientific community acknowledges for a time as supplying the foundation for its further practice. (1962, p.10)

He then elaborates on the nature of these achievements, to which he gives the technical name 'paradigm':

> Aristotle's *Physica,* Ptolemy's *Almagest,* Newton's *Principia* and *Opticks,* Franklin's *Electricity,* Lavoisier's *Chemistry,* and Lyell's *Geology* — these and many other works served for a time implicitly to define the legitimate problems and methods of a research field for succeeding generations of practitioners. They were able to do so because they shared two essential characteristics. Their achievement was sufficiently unprecedented to attract an enduring group of adherents away from competing modes of scientific activity. Simultaneously, it was sufficiently open-ended to leave all sorts of problems for the redefined group of practitioners to resolve.
>
> Achievements that share these two characteristics I shall henceforth call 'paradigms,' a term that relates closely to 'normal science.' By choosing it, I mean to suggest that some accepted examples of scientific practice — examples which include law, theory, application and instrumentation together — provide models from which spring particular coherent traditions of scientific research. *(ibid.)*

A given paradigm thus initiates a period of normal science, which continues until it runs into difficulties that cannot be coped with by following the existing paradigm. At this point there is a paradigm change. Kuhn reviews several examples, after which he concludes:

> To a greater or lesser extent (corresponding to the continuum from the shocking to the anticipated result), the characteristics common to the three examples above are characteristic of all discoveries from which new sorts of phenomena emerge. Those characteristics include: the previous awareness of anomaly, the gradual and simultaneous emergence of both observational and conceptual recognition, and the consequent change of paradigm categories often accompanied by resistance. (p.62)

The resistance which adherents to the old paradigm offer to the new accounts for his use of the term 'Revolution' in the title of his book.

As used by Kuhn, then, a paradigm is a rather specific sort of scientific achievement that serves as a model, an example, for future scientists to follow. But such a model carries with it more than just theories and types of experimentation. It can carry a whole world view, which has to be jettisoned when the paradigm changes. 'View' is used here deliberately. Holding a particular paradigm affects, almost literally, what you do or do not see about the world:

> Can it conceivably be an accident, for example, that Western astronomers first saw changes in the previously immutable heavens during the half-century after Copernicus' new paradigm was first proposed? The Chinese, whose cosmological beliefs did not preclude celestial change, had recorded the appearance of many new stars in the heavens at a much earlier date. Also, even without the aid of a telescope, the Chinese had systematically recorded the appearance of sunspots centuries before these were seen by Galileo and his contemporaries. Nor were sunspots and a new star the only examples of celestial change to emerge in the heavens of Western astronomy immediately after Copernicus. Using traditional instruments, some as simple as a piece of thread, late sixteenth century astronomers

repeatedly discovered that comets wandered at will through the space previously reserved for the immutable planets and stars. (p.116)

Before Kuhn's book, the general picture of science had been one of progressive extension of knowledge into the darkness. Scientists worked away at the coal-face of knowledge, making hypotheses and testing them in order to cut into each new seam of phenomena that emerged. Sometimes progress was gradual, sometimes it came in sudden spurt, but all the time it was a cumulative effort, each worker systematically extending the work done in the past. Once a part of the darkness had been claimed for knowledge, then that area of truth stood valid for all time. Kuhn threw practically every point in this picture into question. By means of a careful analysis from history of what scientists actually did, he showed that the conventional picture was a myth — an ideal view of what scientists had been taught that science ought to be, bearing little relation to the way science was.

First, he showed that the cumulative part of science was not the whole story. 'Normal science' was cumulative, a matter of working out the consequences of an established procedure. But scientific revolutions involved something quite different from building on the work of the past. A past procedure had to be rejected as no longer helpful; there was a period of competition between rivals, rather than cooperative construction; there was a tearing down as well as a building up.

Second, he focused attention on the fact that science was done by real human beings working in real communities. The hallmarks of the successful revolution were defined in social terms: it 'attract[ed] adherents away from competing modes. ...' Certainly the theories had to 'work,' but, for the first time, the idea was raised that the course of science was determined by the social interactions of groups of scientists, as well as by the data they were turning up.

Third, he stressed that the scientific process was not just a matter of making sense of 'data' that arrived automatically from the external world, independently of what theories might be current at the time. Rather, the nature of the data is critically modified by the theories that are in force. Different theories cause one to *see* different things, so that the data automatically reinforce whatever theory is currently in vogue.

3. Beyond the paradigm change

Kuhn's account won widespread appreciation as a reasonable, if oversimplified, expression of the way the major changes in science had taken place. But the questions that it opened up had ramifications which started to destroy science's claim as a repository of absolute truth about reality. The focus of attention for philosophers of science now shifted to the nature of the paradigm change. How in detail did the new paradigm emerge? What was it that inspired one new paradigm rather than another? What was the relation between the new paradigm and the old one?

The conservative view, preserving as much as possible of the idea of accumulating truth, is to say that it is the new facts that determine the new paradigm, and that the new paradigm in some way incorporates the old. After the paradigm change, it might be said, it can be seen how the old paradigm was a partial truth, true up to a point, but extended and enlarged by the wider vision of the new paradigm. The new paradigm has to incorporate the old in the sense that all the data that the old paradigm explained also have to be explained by the new paradigm.

The trouble is, this conservative interpretation hardly fits the facts of what actually happens. We are concerned with reality, and so with the world-view aspect of a paradigm change. And it is usually the case that, when a paradigm changes, the old world-view is not enlarged: it is scrapped completely and replaced by the new. When Copernicus' picture of the sun as the driving centre of the planetary system was adopted, there was no question of it being an extension of the idea that the earth was at the centre; the latter was simply wrong. When quantum theory was accepted, although there was a formal sense in which classical theory was a special case of quantum theory, the world views were so different that it was not possible to hold them together

— a problem that is still with us. And, even if a reconciliation is made later on, when the paradigm change is in process the new paradigm tends to focus on the few troublesome facts that caused the old paradigm to break down, explaining these but offering no adequate account of the many things that the old paradigm might have dealt with quite successfully. The two paradigms operate in different areas of facts, with only limited overlap between them.

These problems were grasped in a dramatic form by Paul Feyerabend in his book *Against Method* (1975), written after, but to some extent independently of, Kuhn's book. His starting point was the idea that, if one proceeded methodically and systematically, then one would continue doing normal science and would never succeed in changing paradigm. Consequently, scientific progress is dependent on the ability of some people to proceed non-methodically. Only by breaking out of the conventionally accepted way of thinking of things can one achieve a paradigm that will account for really new facts; and until one has taken the plunge into the new paradigm, one will not obtain the array of data that will be opened up and which will come to support the new paradigm. At the point of decision, the new paradigm is a leap into a theory which has at that stage almost no support and which probably goes against the facts as they are seen at the time.

Not only does one have to add intuition to the scientific process; one also has to go against all one's normal scientific training to take full account of all the available evidence. Conventional science tends to proceed by what is called *induction:* surveying all the evidence and formulating hypotheses that make sense of it. Paradigm change has to do the opposite. For each rule of normal science, paradigm change requires a counter-rule instructing one to do the opposite.

> To see how this works, let us consider the rule that it is
> 'experience,' or the 'facts,' or 'experimental results' which
> measure the success of our theories, that agreement between
> a theory and the 'data' favours the theory (or leaves the
> situation unchanged) while disagreement endangers it, and
> perhaps even forces us to eliminate it. This rule is an
> important part of all theories of confirmation and cor-
> roboration. It is the essence of empiricism. The 'counter-

rule' corresponding to it advises us to introduce and
elaborate hypotheses which are inconsistent with well-
established theories and/or well-established facts. It advises
us to proceed *counter-inductively*. (1975, p.29)

In expounding this principle, Feyerabend goes considerably beyond
Kuhn. Kuhn takes it for granted that normal science continues until
some fact emerges that makes the old paradigm untenable (he speaks
of the emergence of an anomaly, a contradiction between the old
paradigm and the new facts). Feyerabend takes more seriously the
idea, put forward by Kuhn, that the facts may only emerge once there
is a new paradigm to give a new world-vision. He explains that:

it emerges that the evidence that might refute a theory can
often be unearthed only with the help of an incompatible
alternative: the advice (which goes back to Newton and
which is still very popular today) to use alternatives only
when refutations have already discredited the orthodox
theory puts the cart before the horse. Also, some of the
most important formal properties of a theory are found by
contrast, and not by analysis. A scientist who wishes to
maximize the empirical content of the views he holds and
who wants to understand them as clearly as he possibly can
must therefore introduce other views; that is, he must adopt
a *pluralistic methodology*. He must compare ideas with
other ideas rather than with 'experience' and he must try to
improve rather than discard the views that have failed in the
competition. *(ibid.)*

So Feyerabend is putting forward the view that if counter-induction
is a valuable exercise at moments of paradigm change, then it can
offer valuable insights at all stages of the scientific process. Once we
let go of the idea that science is a repository of eternal truth — and
if we accept the reality of paradigm change, then we must let this idea
go — then the way is open to regard the whole scientific enterprise
as one of allowing a variety of competing theories and pictures to
interact in as diverse a manner as possible. Only through such an

interaction will it be possible to get a genuine critique of the theories; but for this to happen we have to stop regarding any one theory as having a monopoly on truth. This radically alters the relationship between scientific theory and supposed reality:

> Now — how can we possibly examine something we are using all the time? How can we analyse the terms in which we habitually express our most simple and straightforward observations, and reveal their presuppositions? How can we discover the kind of world we presuppose when proceeding as we do?
>
> The answer is clear: we cannot discover it from the *inside*. We need an *external* standard of criticism, we need a set of alternative assumptions or, as these assumptions will be quite general, constituting, as it were, an entire alternative world, *we need a dream-world in order to discover the features of the real world we think we inhabit* (and which may actually be just another dream-world). The first step in our criticism of familiar concepts and procedures, the first step in our criticism of 'facts,' must therefore be an attempt to break the circle. We must invent a new conceptual system that suspends, or clashes with the most carefully established observational results, confounds the most plausible theoretical principles, and introduces perceptions that cannot form part of the existing perceptual world. This step is again counter-inductive. Counter-induction is therefore always reasonable and it has always a chance of success. (p.31)

Where do these alternative paradigms come from, however? The Kuhnian view is that a change of paradigm comes from a response to a problem, posed when the existing paradigms no longer work. Feyerabend, however, holds that the origins of successful alternative paradigms can often be traced back to 'some irrelevant activity, such as playing, which, as a side effect, leads to developments which later on can be interpreted as solutions to unrealized problems.' Moreover, he would not restrict the origins of scientific ideas to the conventional

scientific realm. Recognizing that science is an activity which is part
of the whole matrix of human life, ideas can come from any area into
science and there prove fruitful in some form or other:

> There is no idea, however ancient and absurd, that is not
> capable of improving our knowledge. The whole history of
> thought is absorbed into science and is used for improving
> every single theory. Nor is political interference rejected. It
> may be needed to overcome the chauvinism of science that
> resists alternatives to the status quo. (p.11)

The picture of science that emerges from these works shows that
science as a whole lies in an ambiguous relation to reality, just as we
will find in Chapter 5 when I examine the particular issues of quantum theory and consciousness. Science is not a pure matter of telling
fairy stories; it is an activity that involves a great deal of imaginative
experimentation, in which there is a constant two-way interaction
between the experimentation and the theories. Theories determine our
world view, what phenomena we see (that is, see *as phenomena,* as
potential input to science), and what experiments we ought to do in
order to get the phenomena. The results of experiment in turn
stimulate the growth of theory, creating the soil in which theories
either grow or die, but in a more complex manner than philosophers
of science originally thought. So science is certainly about the external
world, as well as the human world of theory-building.

The human world, however, has a crucial input into the whole
process. The human construction that is science uses as its ingredients
the complex of ideas that are washing around human culture at any
given time, just as science in turn adds to those ideas. There is no way
in which the development of scientific theory can be separated from
the whole cultural and technological fabric of which it is a part. As a
result, the 'world' that the scientist inhabits is neither something
purely dreamed up with no reference to anything external, nor is it a
simple observation of an independent reality. It lives in the relationships between humanity and the external world; and we will see later
(see Chapter 5) that consciousness and the phenomena of quantum
theory are also made up of relationship.

It may be helpful to examine the difference between Kuhn and Feyerabend in terms of what might be called meta-paradigms. All the paradigm changes of Kuhn took place within a fixed belief in a basic scientific method, a basic conception of what was scientifically rational, involving building up observations of difficult facts, using these to formulate hypotheses that might explain the facts (the step called induction), testing the hypotheses, and moving on to gather new facts. Kuhn altered the way philosophers and scientists looked at this process, stressing different aspects of it and introducing the role of paradigms, but not questioning the basic method. This basic method was a kind of meta-paradigm, that had remained fixed for hundreds of years, but which had itself arisen by overthrowing the Aristotelian paradigm in the revolutionary thinking of the seventeenth century.

Feyerabend, in contrast to Kuhn, recognized the provisional nature of the meta-paradigms. The scientific method was not something immutable that could not be questioned. Indeed, it was not even a good account of what scientists actually did. There was a need to bring in different meta-paradigms, as well as different paradigms.

Meta-paradigms are very closely bound up with the whole culture of an age, and Kuhn's analysis of their relative, culture dependent status has repercussions that go well beyond science. It keys in exactly with a modern scepticism about the whole idea of accepted ways of doing and thinking.

4. Postmodernism

The first half of the twentieth century was dominated by a feeling of triumph that science had revealed reality and turned it into a technology that was able to lead humankind into a millennium of plenty, once the outstanding political problems had been solved. The artistic and philosophical expression of this, based on a celebration of the triumphs of modern scientific culture, was termed *modernism*. Progressively, however, other notes crept in. The cynicism and disillusionment engendered by the two European wars and the constant wars in other parts of the world since then; the increasing realization that science had not got the last word on the nature of reality, nor had it got the final solution in terms of technology; the progressive realization that the materialistic culture provided by modern technology and economics gave nothing that answered the deepest needs of humanity for relationships and belonging: out of these and many other trends emerged the new artistic and philosophical ideas that have been termed *postmodernism* — though with the word used in different senses by different people.

If there is any common thread in the diverse movement that goes under this name, it is the desire to expose and reject domination by the world views that have been associated with various different meta-paradigms. The expressions of these meta-paradigms have been called myths, meta-narratives or Great Stories. Examples cited by Dick Hebdige are:

> ... divine revelation, the unfolding Word, the shadowing of History by the Logos, the Enlightenment project, the belief in progress, the belief in Science, modernization, develop-ment, salvation, redemption, the perfectibility of humanity, the transcendence of history through divine intervention, the transcendence of history through the class struggle, Utopia subtitled End of History ... (in Wakefield 1990, p.22)

Just as scientific meta-paradigms define the ground-rules for scientific practice, so these myths define the ground-rules for all rational discourse and behaviour. They are the last court of appeal for saying what is legitimate or illegitimate, and so they claim that they are themselves absolute and not in need of any further justification; either by claiming that they are god-given, or by claiming that the rejection of these ground rules can lead only to chaos, anarchy and other 'self-evidently bad' states. Postmodernism rejects the idea that there can be any such self-authenticating meta-narratives.

When applied to science, this view results in the denial of anything that could be called 'reality.' When applied more generally, the view of postmodernism denies that there is any such thing as 'truth,' in the traditional sense of the term. Hugh Tomlinson expresses the traditional view as follows:

REALITY AND BEYOND   35

> The 'commonsense realist' sees the world as being
> objectively ordered independently of all human activity.
> Science seeks to provide theories which 'mirror' this
> objective ordering, theories which are, in a word, 'true.'
>   The basic 'given' of commonsense realism is unob-
> jectionable: the world is relatively independent of our
> dealings with it. But this is elevated into an ontological
> thesis that there is a single, objectively structured reality
> independent of human thought and actions. According to the
> realist, 'the world' consists of some fixed totality of mind
> independent objects. There is exactly one true and complete
> description of 'the way the world is.' Truth involves some
> sort of correspondence between words or thought-signs and
> external things and sets of things. It follows that the key
> feature of language is 'reference': bits of language are used
> to 'refer' to bits of the world. A statement is true if it
> successfully refers to the world. Whatever our subjective
> perceptions, the world is objectively 'out there' and the aim
> of our theorizing is to provide a theory which 'copies' or
> 'corresponds' to it. (in Lawson and Appignanesi 1989, p.45)

Tomlinson agrees that this is a reasonable approach to questions such
as whether 'there is food in the fridge': to decide this, you can go and
look in the fridge. But what seems at face value quite reasonable starts
to break down when it is extended to modern science:

> The commonsense realist wants to expand this picture into a
> general account of the relation between words and world, a
> general theory of truth. There are immediate difficulties with
> this move. What seems obvious in relation to ordinary phy-
> sical objects such as cars and houses is difficult to apply in
> relation to the objects of the physicist or astronomer. There
> is no straightforward and clear sense in which talk about
> quarks or quasars 'corresponds' to the world. The
> acceptability or utility of such talk depends on complex
> procedures of observation and calculation which are entirely
> unknown to most of us. *(ibid.)*

Different postmodern writers react to this in different ways. All would agree that we have to drop the idea, in R. Rorty's words, that 'enquiry is destined to converge to a single point — that Truth is "out there" waiting for human beings to arrive at it.' But some would say that there is, none the less, a pragmatic sense to the word 'truth' that allows one to use the word, but in a more restricted way than traditionally. Rorty describes this as:

> the ethnocentric view that there is nothing to be said about either truth or rationality apart from descriptions of the familiar procedures of justification which a given society — ours — uses in one or another area of enquiry. (in Lawson and Appignanesi 1989, p.11)

He claims that, if one adopts this limited notion of truth, then one can indeed support the idea that society is progressing — though not *towards* some absolute goal:

> Paul Feyerabend is right in suggesting that we should discard the metaphor of enquiry, and human activity generally, as converging rather than proliferating, becoming more unified rather than more diverse. On the contrary, we should relish the thought that the sciences as well as the arts will always provide a spectacle of fierce competition between alternative theories, movements and schools. The end of human activity is not rest, but rather richer and better human activity. We should think of human progress as making it possible for human beings to do more interesting things and be more interesting people ... we should think of 'true' as a word which applies to those beliefs upon which we are able to agree, as roughly synonymous with 'justified.' To say that beliefs can be agreed on without being true is ... merely to say that somebody might come up with a better idea. (p.14)

At first sight this seems to be a rather elegant and enlightened compromise. Although one gives up the idea of an absolute goal (an idea which, though comforting, was always rather theoretical), one can still get on with the business of science by operating within whatever pragmatic criteria one's fellow scientists have agreed on, and at the same time feel that one is building a better and more entertaining society. There are, however, a number of problems. One lies in the phrase: 'beliefs upon which we are able to agree.' Who are 'we'? As Tomlinson says: 'our society offers not one, but a whole multitude of "rationalities," of the "forms of justification".' Is Rorty saying that progress consists of whatever the dominant group of intellectuals decide to agree on? This is a course that could lead to a most pernicious form of dictatorship.

5. Morality and feminism

Here the argument, which started in an examination of the structure of science, has clearly started to have moral implications. This is made quite explicit by many postmodern authors, many of whom are much more concerned with issues of the meaning of right and wrong and the way in which these are embodied in the legal structure of a society, than with the structure of science. I have already touched on moral questions in describing the way in which the scientist's dominant position as an arbitrator of reality is appealed to when scientists are called as expert witnesses in our legal system. So it will be helpful at this stage to address the moral dimension.

Two points are central to the place of morality in our arguments. First, the idea that it is possible to separate morality from the nature of physical reality is itself a peculiarity of the modern scientific view of reality. It depends on the separation of reality into the two separate layers of the physical/scientific and the human, with few direct links between them. For non-scientific cultures, on the other hand, morality is inextricably bound up with the whole of the world. Myths about creation are told, not out of academic interest but in order that the hearers might conduct themselves correctly in society and have the correct behaviour towards the world. If we are to question the scientific picture of reality then we will inevitably be involved in morality.

Second, the assertion that morality is separate from the physical world is itself a moral position. It is a very specific prescription about how one is to determine moral questions: namely, one is to ignore the physical world and look only at the human implications. This has immediate implications for the morality of our actions. It legitimizes any destruction of the environment if that destruction can be justified on the basis of human good, for example. If I were to ignore morality in this book, then by my silence I would be arguing a particular moral position.

Because of these two points, any criticism of the modern scientific view of reality immediately brings moral questions into the arena. At the same time, however, the postmodern criticism raises a question mark over the whole area of morality, because the same arguments that it applied to science can also be applied to morality. Many postmodernists would argue that, just as there is no hotline to physical truth, science becoming an activity in which many competing models are jostling for acceptance by society, so there is no hotline to absolute moral truth. Every section of society has its own moral code, and there is no final court of appeal at which we can compare them. This entry of morality into the argument thus raises the stakes in the whole discussion, and opens up a dark prospect of complete moral relativism, a world in which anything goes. What is going on here? We begin by treading a path that seems to liberate thinking from the arid despair of the mechanical scientific picture of the universe; we end in the equally arid desert of a world with no morality.

It is vital to appreciate that this is not inevitable. It is all too easy to polarize the debate into a liberal-versus-conservative dualism: either one clings to eighteenth century science and eighteenth century morality, or one throws out all science and all morality. I am not advocating throwing out science, but enlarging it; I am not advocating throwing out morality, but rooting it in a deeper concept of reality. To escape from this dualism, to understand how to criticize scientific reality without at the same time buying-in to a programme of relativism in all areas, we need to take account of a historical process that has been driving both the development of modern science and also one section of the postmodern movement, a process that has only recently been brought under analysis by feminist writers: the patriarchy.

Literally, the term patriarchy means the 'rule of fathers' in the political sense, the system dominant in a large part of the world for several thousand years in which the official structure of society has been determined by men. As used by feminists, however, the term means not just the legal rule of men, but the whole system accompanying it in which men have tried to monopolize power in all areas of society, having political, economic, sexual and intellectual dominance.

Charlene Spretnak, in particular, has applied feminist analysis to postmodernism (1991). She points out that the banner of postmodernism brings together many different groups of people, united in their rejection of any fixed meta-paradigm (meta-narrative) labelled 'reality,' but following many different agendas. Some of these groups, though they may criticize the absolutist position of modern science, are in fact following exactly the same patriarchal agenda as the founders of modern science, and it is this, rather than their approach to reality, that leads them to disengage from moral commitment, just as the founders of science separated their view of reality from morality. So, at a deeper level, many of the postmodernists are working out exactly the same programme as the scientists who introduced atomism in the seventeenth century.

As I shall describe briefly in Chapter 2, the driving force behind atomism, and hence behind the whole early development of science, may well have been a desire to dominate and control nature. This desire drew its strength from the political process of building societies in which the men dominated and controlled the women, the latter being associated with nature. As has now been well documented by Spretnak, Roszak and others, this linking of the two dominations instituted a vicious circle: once men adopted a position of domination over nature, they felt alienated from their roots in nature, unable to respond to nature's actions, and at the mercy of a nature that came to be perceived as hostile. The fear and hostility towards nature could be discharged by projecting it onto woman, so that the successful dominance of women became a necessary defence mechanism for the fear and alienation caused by taking up a dominating position towards nature.

It was a system that 'worked' in material terms (until the time of the current ecological crisis), but at a deep psychological and social cost. Once men, the dominant voices in society, had moved to a position of regarding nature and women as hostile, the basis of all human relationships was undermined: relationships with the material world, requiring an intuitive, sympathetic response to the world of nature; and human relationships which called for this same response to the totality, physical and spiritual, of the other person. Of course, such relations continued — human society would instantly disintegrate without them — but they continued in a form distorted by and in contradiction to the basic norms of patriarchal society. The dominant norm became that of the isolated hero, performing his exploits in splendid isolation, or the solitary genius giving birth to ideas in a lonely attic. It was, and remains so today, a norm of 'every man for himself' (and never mind about the women).

It is striking how many aspects of postmodernism, far from being a bare denial of norms and meta-narratives, are in fact a covert reinforcing of the norm of every man for himself. Once reality is denied, it is impossible to have an intimate relationship with it; all that is left is love for a spectre. The concepts of love and commitment are condemned by many postmodernists as enslavements to meta-narratives that threaten the freedom of the individual to determine his or her own story. The best one can achieve is to behave pragmatically *as if* the beloved were real. George Steiner, in *Real Presences* (1989) has characterized postmodernism as the denial of the actual presence of anything outside the individual, an individual locked inside his or her own language game. This is not a neutral stand, but a very specific affirmation of the patriarchal programme of refusing to enter into any relationship with external physical entities, other than a relationship of control and domination.

This analysis thus brings with it a completely new understanding of the choices before us as we consider the nature of reality. A critical examination of the historical processes of science is only the first step. The real choice is between the separateness of the patriarchal man, who regards every form of reality, whatever it may be, as 'out there,' external to himself, not demanding commitment; and the engagement of the post-patriarchal person with the world, in a committed relationship with whatever presents itself. The second way is a path that does indeed question particular moral codes, but it involves a human commitment at a moral level. This path leads not to a moral desert, but first to a concept of reality lying in the relationship of commitment, rather than an external concept of reality, and finally, as we shall see in the last chapter, to a reality that is beyond concepts.

Spretnak (1991) describes well how this commitment works in discriminating between those postmodernists who maintain their relatedness to the world, and those who are leading us into a moral desert. The second sort of person is removed from real presence and real relationship. When this happens, oppression and violence become legitimized, or at least are freed from any categorical restraint. To discern this requires a commitment of the whole person, body and mind, a critical awareness of one's deepest moral instincts.

6. Reality and duality

Thus we return to the question that I posed to the quotation from Rorty: when reality becomes 'beliefs on which we are able to agree,' who are 'we'?

And yet we know the extent to which society has been paralysed and corrupted by clinging to myths about the nature of reality, and we know how science has often refused to let go of long outdated paradigms. The key task facing science and society today is to break free of the dualism that asserts that either reality is a particular construction proposed by science, or there is no such thing as reality at all. As we shall see again later, through a study of quantum theory, reality is; and yet it is not any one thing that can be encapsulated and grasped. Reality is to be found within relationships, not within concepts.

So we can agree with Feyerabend and the postmodernists that we can never accept any one myth, and declare that this is the final reality. But that does not commit us to the extreme postmodernist position of denying the value of the myth-making process itself. Making myths, building meta-paradigms, is a vital tool in guiding us into forming those relationships with each other and with the physical world on which our whole being depends. Humanity is in need, as never before, of grasping the dynamic process described by Feyerabend — not as a denial of the real presence of the world, but as the means whereby we can continually and creatively engage with that presence.

Chapter 2
The Leap into the Void

I have argued that our view of reality has been shaped by the historical process taken by modern science. The first revolution in scientific thought, in the seventeenth century, founded modern science and with it the dominant view of reality today. As I described in the previous chapter, reality, for most people, has come to mean scientific reality (even if they are unaware of it). Since that revolution we have been living under the domination of seventeenth century ideas, a period of three hundred years in which a particular view of the world was penetrating every aspect of life and sinking deeply into every recess of human thought. Protests and criticisms were raised (we shall be referring to Goethe, Blake and others), but these authors were variously regarded as visionaries, romantics or (in some quarters) cranks. Apart from this, the ideas of the seventeenth century (in their own time the radical speculations of a handful of revolutionaries) have been simply 'the way the world is.' To understand our situation we have first to free ourselves from taking for granted what the seventeenth century scientists had proposed. And to do this, we need to go back and understand the nature of the revolution they initiated.

1. The classical inheritance

Prior to the seventeenth (or late sixteenth) century, Western (and earlier on, Muslim) thought had been dominated by the philosophy of scholasticism, derived from Aristotle. It is hard to overestimate the importance of this genius, many of whose insights — into, for example, the nature of space and time — are only now starting to find their place in scientific thought. His starting point, the starting point of all philosophy up to the modern era, was the world that we are aware of in everyday life, a world of stars, sky, sea, animals, colours, sounds, emotions and the rest. Aristotle recognized that there was a great mass of folklore about the world, usually regarded as 'common sense,' and also a great many theories proposed by previous thinkers who speculated, often wildly, about the world: that it was ultimately all made of water, as Thales had proposed; that it was based on number, as the followers of Pythagoras held, and so on. Throughout his books, he usually started by looking critically both at these theories and at the 'common sense' of folklore, rejecting those elements that either were illogical or did not fit the facts, founding his thinking on a careful view of what the world seemed like to his contemporaries.

So for him 'reality' kept closely to experience — but experience analysed by the keenest logical thinker who has ever lived. In retrospect, therefore, we can see his world as a combination of the experience of people of ancient Greece with the logical structures embodied in the Greek language that Aristotle used. For, just as he accepted without question the equal reality of everything that we perceive, so he accepted without question that the right way to handle them was by talking about them carefully in the Greek language.

Much later his system was taken up in the West, translated into Latin and developed into a huge and elaborate intellectual structure. Aristotle's view of the universe was merged with the Christian view to form an integrated intellectual construction. As Wildiers describes it:

> For the medieval theologian, then, the world was a perfectly ordered whole. Moreover, God's wisdom required that he create order in everything he did. ... The world was also defined as an ordered collection of creatures. To doubt the reality of a perfect world order approached blasphemy. If ever there was any uncertainty whether a particular being existed, it was the criterion of world order that settled the issue. Thus wisdom compelled God to create the angels, without whom the world order would not have been complete. What greater joy or perfection can there be than to assimilate this perfect world order and to have a thorough knowledge of it? There was no doubt that this order is

immutable: it dates from the creation of the world and will hold good until the end of time and beyond this into eternity ...

Besides being perfect and immutable, this world order was also considered hierarchical. What is at the top is naturally nobler, and what is nobler is better and more powerful, and whatever is better and more powerful exercises the greatest influence. (1982, p.57f)

These conceptions led to the idea of the universe as a series of concentric crystal spheres, with the earth in the centre, then the moon, the planets and the sun, then the stars, and outside this the orders of angels.

In addition to this global picture, the medieval world inherited Aristotle's logical approach to the way the world worked, and in particular the way one thing turned into another: the acorn into the oak, water into ice or steam, boy into man into dust and so on. The universe was governed by a web of changes in which each object could have a potentiality for being something different, or being somewhere different, and in which this potentiality could become actual when the external influences were right.

In Aristotle's hands, this system was a wonderfully subtle tool for uncovering the processes that made the world tick. In the hands of the medieval scholars, however, it became an arid set of rules divorced from experience. The valuable insights of Aristotle's original work were submerged under layers of verbiage, until the whole system became completely bogged down under the weight of its own words. It came to the point where the only way forward was to scrap the whole thing and start again, with something so fresh and different that it could break free of the clutches of past thinking.

The Greek world had passed on to the West other ideas that contributed to this new start, one being the very old notion of different levels of reality. Plato and his followers, for example, had argued that what comes through our senses is 'appearance,' not reality; whereas the true reality resides in a separate realm, accessible by the intellect and not the senses. In one of the most famous passages in the whole of philosophy (to which we shall return in Chapter 8), he likened our view of the world to that of people inside a cave, watching shadows cast on the rear wall. We mistake the shadows for reality; worse, we habitually sit with our back to reality so that we are incapable of looking out of the cave to see it. For Plato the task of reaching reality is to turn around our habitual mind-set so that, instead of fixing on appearances, we fix our interior gaze on the eternal ideas that are true reality.

2. The patriarchal process

Later I shall describe how these classical ideas were overturned in the seventeenth century to give the foundation of the modern view of reality (see p.50). Until recently this dramatic revolution seemed arbitrary and without cause. Why did it happen then? Was there some ataclysmic historical event that caused this revolution in thought? Only now are we becoming aware of the historical context of the scientific 'revolution' through the work of ethnologists and feminist philosophers: the answer to these questions is probably to be found in the feminist approach to postmodernism that I sketched earlier (see p.37). From this point of view, there is no sudden event that transforms society's ideas. Rather, the classical ideas of Aristotle should be seen as part of a continuous development which finds its culmination in the transformation of these ideas in the seventeenth century.

The primary process involved is the successive transitions that have taken place in Western society over the ages. First, in prehistoric times, from a hunter-gatherer society, through a horticultural one, to an agricultural one; then, overlapping into historical times, into an urban society. Felicitas Goodman has acutely analysed the way in which this has resulted in a progressive alienation of humanity from nature, so that the city culture is characterized by an increasing stress on dominating and controlling the environment, rather than living harmoniously with it:

> The change ushered in by the horticulturalists was profound.
> ... *humans began aggressively asserting control over the*

habitat. By working the soil they forced Mother Earth to yield what she did not readily have available. This represented a fundamental break with hunter-gatherer attitudes. While no doubt the cultivation of plants was a solution in a period of great need, what was experienced as a rape of the Mother still produced a massive feeling of guilt. It is with the horticulturalists that we encounter for the first time a prediction about the end of the world as ultimate retribution for this sacrilege. (Goodman 1988, p.19)

With the move from horticulture to agriculture, the estrangement from the earth proceeds further, so that now there is no longer any feeling of guilt for what has happened:

The earth is seen mainly as a docile partner; no feeling of guilt attaches to despoiling it. ... Instead of being plagued by guilt feelings towards the earth, agriculturalists are beset by paranoia toward 'bad' plants — weeds — and toward equally 'bad' animals that intrude into their fields, all of which are to be exterminated in order to protect the harvest. These attitudes are part of a hidden, more abstract complex. In the agricultural adaptation, humans have drawn a magic circle around themselves, their homes and their fields. They are confined within their fences, but also protected against what lies beyond. Any intrusion is experienced as threatening in the extreme. The intruder brings with it danger, sickness, pollution, the curse. (p.26f)

Finally this process of alienation from the Earth is completed in the city-dwellers, the urbanites, to whose culture most of us belong:

Urbanites are divorced from the habitat. The earth, the sky, the rain, the plants, and the animals are not their partners in the struggle for subsistence. The alienation becomes even more pronounced with the advent of industrialization, eventually causing the destruction of traditional peasantries and the takeover by agribusiness. The latter has no direct

link with the habitat at all. Rather, it exploits it by the
interposition of machinery. (p.29)

At the same time this process involved the progressive introduction of
ideas of property and personal appropriation of resources, resulting in
women becoming the property of the men and being relegated to an
inferior status. During the rise of city-based culture ('civilization' in
the literal sense of the word) this tended to take the form of a male
city-culture dominating a female culture that still retained some of the
older contact with the natural world, giving rise to the patriarchy that
we examined above (see p.39).

This means that the process of alienation from nature is precisely
paralleled by the growth of the patriarchy. It is this realization of the
historical depth of the patriarchy, and its link with alienation, that
constitutes one of the major revelations of today. As Roszak puts it:

> The root of the problem lies in prehistoric sociology.
> Modern institutions, whether socialist or capitalist, have
> been corrupted to their core by patriarchal values that vastly
> predate the modern property system and all other class
> relations; it is the prototypical form of social domination. In
> order to find a culture free of its destroying influence, one
> may have to reach back in history to that early Neolithic,
> village-based, agrarian culture that the archaeologist Marija
> Gimbutas calls 'Old Europe' ... in [her] influential
> formulation, it was the Indo-European incursion of warlike
> Indo-European nomadic tribes, worshippers of masculine
> sky-gods, that replaced the matricentric cultures of Old
> Europe with an 'androcratic warrior' society that henceforth
> claimed all the virtues of 'civilization' for itself. (1992,
> p.236f)

And he too describes incisively the nature-woman link that underlies
this history:

> Gender identity in all but a few exceptional cultures of the
> world stems from the assumption that 'women [in Sherry

Ortner's classic formulation] are to nature as man is to
culture.'... The traditional imagery of Mother Earth and
Dame Nature is far from a mere poetic simile. It embodies
an age-old association between women and the natural
world that has transformed both into alien and contemptible
targets of male domination. *(ibid.)*

The driving force behind the seventeenth century scientific revolution
was this growth in the ethos of masculine control over nature. Now
there is one very significant indicator of the way in which the growth
of this ethos came to a head in Europe precisely at the time that we
are examining: namely, the burning of witches. What happened was
that the oppression of women/nature was fanned by medieval Christianity into a paranoiac hatred. The original teaching of Christianity
was overlaid, as we saw earlier (see p.44), with a hierarchical conception of the cosmos in which man, woman and nature formed a
descending pecking order. As a result, the older religions, which gave
prominence to the wisdom of women as mediators of the power of
nature, were regarded as a direct threat both to the patriarchal social
order and to the established divine order of the universe (the latter
being simply a projection of the former).

In the sixteenth century this growing confrontation between men
and women/nature reached its ultimate expression in the witch-
burnings that swept across Europe, a 'final solution' (linked in many
ways to the later Nazi holocaust) in which any woman who had any
contact with the ways of nature, or who was not willing to be
subservient to the masculine culture, was branded as a witch and burnt
in a purge that cost nine million lives. The scientific revolution thus
occurred at the peak of the bloody triumph of the dominance of nature
by man. The new science, based on the theory of atomism, was
grasped as a means of obtaining complete control over every aspect
of nature: once the workings of the atoms had been elucidated, there
would be nothing that was unknown, nothing that could not be
controlled.

3. The seventeenth century

Koestler entitled his biography of the astronomer Kepler *The Watershed,* indicating the way in which his life, from 1571–1630, spanned the transition from the final stages of medieval science to the first beginnings of modern science. At this time the air of confidence among the European ruling classes, the progress of the economy and technology, reached the point where a reverence for the wisdom of the past was being outweighed by a belief in the new possibilities of the future. Confidence had reached the stage where one could start to think out afresh the way the world was, using the new insights given by a world that was now very different from that of ancient Greece.

One seed of the new thought was Plato's distinction between appearance and reality. At any rate, the idea that 'all might not be as it seems' was not new at the time when a few thinkers were brave enough to make this the cornerstone of a new approach to science and philosophy, letting go of all the security of their common sense and previous thought, and leaping into the void of speculation which Aristotle had been so careful to criticize. Like Plato, they made a sharp distinction between appearance and reality. But, whereas the followers of Plato would say that all external objects are only semi-real, authentic reality residing in the intellectual realm of ideas, the new thinkers placed reality not in a world of heavenly abstract ideas, but in the scientific concept of matter.

A good way to understand the new thinking is by looking at the change that took place in the way in which people thought about human sense-organs (smell, vision and so on). Before this time, the prevailing view was that these organs *discovered* qualities that were present in the world. Some things are red, and are revealed to be red by our eyes; some foods are sour, and our taste reveals this to us. Now the idea emerged that the sense organs *created* these sensations of redness, sourness and so on, out of something in nature that in itself had none of these qualities.

Thus in 1623 Galileo was writing that:

> I think that tastes, odours, colours, and so on are no more
> than mere names so far as the object in which we place
> them is concerned, and that they reside only in
> consciousness. Hence, if the living creature were removed,
> all these qualities would be wiped away and annihilated.

A little later (perhaps around 1630) Descartes was explaining that
there could well be a difference between actual things and the
perceptions that we have of them. In the case of light, for example:

> the first thing I want to make clear to you is that there can
> be a difference between our sensation of light ... and what is
> in the objects that produces that sensation in us. ... For,
> even though everyone is commonly persuaded that the ideas
> that are the objects of our thought are wholly like the
> objects from which they proceed, nevertheless I can see no
> reasoning that assures us that this is the case.

This is a radical change of view. Instead of the eyes being regarded
as windows through which the world enters more or less directly into
our minds, they now become complex instruments that convert something physical (the light from an object) into something mental — the
idea, or image, of that object. Although at this stage scientists had
very little idea about what light was, how the eyes operated, or what
the brain did, it was clear that in fact the eyes were not mere
windows; that some sort of complicated transmission process was
involved between the light and the brain, even though its nature was
unknown; and hence that the sensation of which we are aware might
not be a simple copy of the real thing, but could be a construction of
this complex process of transmission.

As a result, it came to be thought that 'appearance' involves all
sorts of qualities (the redness, sourness and so on which later philosophers, following Galileo and Descartes, were to call *secondary
qualities)* only indirectly associated with objects in themselves,
according to these writers, which are added by mechanisms in our
brain and sense organs.

This now starts to make a radical change in our view of reality. If, as Aristotle assumed, things on the whole are as they appear to be, then reality is just what the world seems to be. If I see a red flower, then, on the old view, if I am in good health and the light is good, then there really is a red flower, and its redness and its flowerness are part of this reality. On the other hand, if the redness is something added by eyes/brain, then the redness is not part of the reality of what of what I am seeing. It is as much an illusion as the pink elephant I might see when I am not in the best of health.

Once one goes down this path, however, our whole concept of reality starts to be questioned. As soon as I start doubting my basic senses, there seems to be no foundation for understanding anything. If, as Galileo held, 'tastes, odours, colours, and so on' are unreal, what is real? Is the 'reality' that lies behind my perception of the red flower a colourless flower to which I add the colour red, or is it some totally abstract 'thing' to which I add the concept of flowerness? Are the objects that we see around us real at all? So a new notion of reality starts to be formulated, with its principal architect being Descartes. Like Plato's concept of reality as the world of ideas, the new concept of reality placed a big stress on the intellect, but it also emphasized the importance of the material, physical world.

The key to finding a new concept of reality was to distinguish between things in appearance that we could rely on and things that were added by our mental processes. This was one of Descartes' main concerns and his conclusion was that we could rely on certain aspects of the world:

> It follows that corporeal things exist. They may not all exist
> in a way that exactly corresponds with my sensory grasp of
> them, for in many cases the grasp of the senses is very
> obscure and confused; but indeed, everything we clearly and
> distinctly understand is in them, that is, everything,
> generally speaking, which is included in the object of pure
> mathematics. (AT VII 80)

There are certainly echoes here of Plato's view that 'appearance' can be deceptive, but that the intellect can reach a higher, more certain reality, particularly in the way that Descartes singles out mathematics as the key to extracting what is real in the world. For him, physical reality consists of geometry: positions, lengths, distances, angles. Practically everything else is added by our own brains/minds. The decisive point, however, is that Aristotle has been dramatically overthrown. Almost everything that he talked about has been relegated to an invention of the human mind, and in place of wordy discussion in the Greek or Latin language we have the logic of mathematics (which, at the time of Descartes, was undergoing a transformation into the abstract symbolic system that we know today).

If, now, so many properties and sensations are being regarded as pure products of the human mind, it is natural to ask how this happens. What is this 'mind' that is responsible for so much of what we experience? Descartes' answer leads into another major ingredient in his philosophy: dualism. Almost all cultures — with the exception of Judaism and, following on from it, very early Christianity — have, to a greater or lesser extent, believed in two quite different and separate kinds of things making up the universe: bodies and souls. Descartes adopted this view, holding that body was essentially geometrical, characterized by extension, while the soul had no extension and was responsible for the higher mental processes. Many lower mental processes are dealt with by the brain, part of the body, such as:

> waking and sleeping, the reception by the external sense
> organs of light, sounds, smells, tastes, heat and other such
> qualities, the imprinting of ideas of these qualities in the
> organ of the 'common' sense and the imagination, the
> retention or stamping of these ideas in the memory, the
> internal movements of the appetites and passions, and
> finally the external movements of the limbs which aptly
> follow both the actions and objects presented to the senses
> and also the passions and impressions found in the memory.
> (AT XI 120)

This account, which sounds very modern in the way it attributes so much to the mechanical action of the brain, makes it clear that it is the brain, not the soul, which is responsible for the qualities of colour and the rest that Galileo banished from the physical world. The soul is responsible only for reasoning and the formation of meaningful speech, purely intellectual activities.

The seventeenth century revolution thus involved the relegation of qualities to the human brain; the adoption of dualism; and, finally, atomism. This was a theory propounded by Greek writers (notably Epicurus and Democritus) and described in a later Latin poem by Lucretius. It had two basic principles. First, the basic structure of the universe was to be thought of as an infinite void in which moved vast numbers of minute particles which repeatedly collided with each other, producing all physical processes by their collisions. Second, these particles are unchangeable, indivisible and indestructible. Epicurus regarded the two principles as closely interlinked: there would be little future in a universe built on a foundation of particles if the particles themselves were gradually worn away in the course of their collisions. He wrote:

> Of bodies, some are composite, others the elements of
> which these composite bodies are made. These elements are
> indivisible and unchangeable, and necessarily so, if things
> are not all to be destroyed and pass into non-existence, but
> are strong enough to endure when the composite bodies are
> broken up, because they possess a solid nature and are
> incapable of being anywhere or anyhow dissolved.

Lucretius set out the theory in 7400 lines of eloquently lyrical Latin verse, dedicated to an unknown person named Memmius. And, in case the recipient remains unconvinced by this, there is much more where it came from:

> But if you are slack or shrink a little from my theme, this I
> can promise you, Memmius, on my word: so surely will my
> sweet tongue pour forth to you bounteous draughts from the
> deep wellsprings out of the treasures on my heart, that I fear
> lest sluggish age creep over our limbs and loosen within us
> the fastenings of life, before that the whole store of proofs

> on one single theme be launched in my verses into your
> ears. (I 410–417)

Clearly, he is the sort of author that every editor dreads. However,
after this passage he comes to the point:

> But now, to weave my task again into words, all nature
> then, as it is of itself, is built of these two things: for there
> are bodies and the void, in which they are placed and where
> they move hither and thither.

The void is infinite; because if it were finite it would have to have a
boundary, and the idea of a void with a boundary is absurd:

> If one were to run on to the end, to its farthest coasts, and
> throw a flying dart, would you have it that that dart, hurled
> with might and main, goes on whither it is sped and flies
> afar, or do you think that something checks and bars its
> way? For one or the other you must needs admit and
> choose. Yet both shut off your escape and constrain you to
> grant that the universe spreads out free from limit. For
> whether there is something to check it and bring it about
> that it arrives not whither it was sped, nor plants itself in the
> goal, or whether it fares forward, it set not forth from the
> end. In this way I press on, and wherever you shall set the
> farthest coasts, I shall ask what then becomes of the dart. It
> will come to pass that nowhere can a bound be set, and
> room for flight ever prolongs the chance of flight.
> (I 970–984).

Within this infinite void the atoms are in constant turbulent motion:

> For when as [the atoms] move, again and again, they have
> met and clashed together, it comes to pass that they leap
> asunder at once this way and that ... And the more you
> perceive all the bodies of matter tossing about, bring it to
> mind that there is no lowest point in the whole universe, nor

have the first-bodies [atoms] any place where they may
come to rest, since space is without bound or limit, as I
have shown in many words [!], and it has been proved by
true reasoning, that it spreads out immeasurable towards
every quarter everywhere. And since that is certain, no rest,
we may be sure, is allowed to the first-bodies moving
through the deep void ... (II 85–96)

This picture of a strange and restless infinite void, a picture that burst
upon the thinkers of the seventeenth century with the bracing effect
(so I am told) of a cold shower, was explicitly linked by Lucretius to
the other plank of the new thinking: the idea of secondary qualities
arising in the brain, with only shape being real. He regarded colours,
for example, as the result of the way in which atoms of different
shape interact with the eye:

And since the pupil of the eye receives in itself a certain
kind of blow, when it is said to perceive white colour, and
another again, when it perceives black and the rest ... you
may know that the [atoms] have no need of colours, but by
their diverse forms produce diverse kinds of touch. (II
810–815)

These Greek ideas were revived in Europe by Gassendi early in the
seventeenth century, and at first were received cautiously. Descartes
accepted the idea that nature was made of particles, but objected to the
word 'indivisible.' He wrote:

We know that there can be no atoms, that is, parts of matter
by their nature indivisible. For if there were such things,
they would necessarily have to be extended, however small
we imagine them to be, and hence we could in our thought
divide each of them into two or more smaller ones, and thus
we could know that they are divisible. For we cannot divide
anything in thought without by this very fact knowing that
they are divisible.

Soon, however, others accepted the idea enthusiastically, regarding Descartes' argument as a purely verbal objection. Even if atoms were, in some sense, in principle divisible, the important thing was that in practice they could not be divided. Thus by the middle of the century there was in place the three ingredients of the new view of reality: secondary qualities, duality, and atomism, with atomism resting on two planks — the assumption of atoms and the void.

The adoption of this world view illustrates well the principles of scientific change due to Feyerabend and the postmodernists already mentioned (see p.29ff), in contrast to the earlier view — due to Popper and others — of science as a progressive unfolding of reality. On the earlier view, science was presented as moving forward by gathering evidence and then formulating hypotheses on the basis of the evidence. The atomic theory, however, was definitely not the result of the accumulation of scientific evidence in its favour. There would be not a shred of evidence for atomism for another two hundred years; it seemed to fly in the face of all common sense — the universe gave no impression of being made up of particles and the void, but on the contrary seemed to be filled with extended substances such as earth and air all touching each other and filling space, in the way described by Aristotle.

On the earlier concept of science, the adoption of atomism had been one of the most inexplicable events in the history of science. Just as Feyerabend points out, at first nothing was explained by the atomic hypothesis and much that had been explained became inexplicable. The reason for its adoption was not the weight of evidence, and we therefore need to look elsewhere for this reason — for instance, to the patriarchal process described above (see p.39).

The success of the idea, once adopted, was due not to its plausibility as a theory, but because it forced people to look at the universe in a new way. From now on the universe was regarded as a giant machine, and its working was to be understood by looking at its smallest parts, not at how it functioned as a whole. Atomism was the perfect accompaniment to Descartes' picture of a physical, geometrical universe devoid of qualities, because it made this universe into a machine. 'Reality' was transformed. The leap into the void of the new philosophy now included the literal void in which the atoms moved, which I shall examine next.

4. Atomism after the seventeenth century

I shall be referring to 'atomism' as this general physical view of the world as made up of particles and the void. It was a view that was central in motivating Newton to produce his definitive account of how the machine-world worked, and, ever since, atomism has been associated with Newton's mathematical laws of the universe. Of course, the word 'atom' now means something very different — a particular structure of electrons and a nucleus that forms the basis of chemical elements. Though the word has changed its meaning, the image of atomism still lives on.

Eventually evidence did appear for the hypothesis of atomism. One such piece of evidence was signalled in advance. The one apparent exception to the general lack of evidence was a phenomenon noted by Lucretius:

> For look closely, whenever rays are let in and pour the
> sun's light through the dark places in houses: for you will
> see many tiny bodies mingle in many ways all through the
> empty space right in the light of the rays, and as though in
> some everlasting strife wage war and battle, struggling troop
> against troop, nor ever crying a halt, harried with constant
> meetings and partings; so that you may guess from this
> what it means that the [atoms] are for ever tossing in the
> great void. ... such jostlings hint that there are movements
> of matter too beneath them, secret and unseen. ... you may
> know that this shifting movement comes to them all from
> the [atoms]. (II 114–132)

This observation was the source of much later controversy. To cut a long story short, it emerged that the motion of these dust particles was due to convection currents, and not to the buffetings of atoms; but that there was another very similar effect explained precisely as indicated by Lucretius. This was Brownian motion, the constant jiggling motion of particles of pollen suspended in water. When its theory was finally worked out by Einstein, it was realized that this was indeed direct evidence of atoms, and could be used to estimate the sizes of atoms.

Though less direct, the central evidence for the idea of atomism came from the theory of heat and the behaviour of gasses. According to this, heat was always an expression of the movement of the elementary particles making up any substance (particles which by this time, in the eighteenth and nineteenth centuries, were known as molecules rather than atoms). In solids this motion was a vibration around a fixed place; in a gas it was a Lucretius-like motion through the void with occasional collisions. The great achievement of the nineteenth century physicists, culminating in the work of Boltzmann, was to show how the detailed behaviour of gases — the way their pressure varied with temperature, for instance — could be explained in this way.

As already noted, a quite separate strand of evidence came from chemistry, where the combination of atoms was postulated as a means of explaining the way in which a given chemical was always made up of exactly the same proportions of more elementary substances. As a result, the chemists postulated the existence of a large number of different atoms (the number eventually rose to 92, and has increased further in modern times) which combined to form molecules, the basic particles of chemical compounds.

Alongside these scientific successes, we need to set the historical origins of atomism. I described earlier how the new science constituted a means of obtaining complete control over every aspect of nature, control that would be absolute once the workings of the atoms had been elucidated in complete detail. This vision was played out in detail over the succeeding centuries. It is the dream that was dramatized by Goethe in *Faust,* the story of the man who sold his soul to the devil in order to gain complete control over nature, and it is the dream that still powers science, with its billions of dollars and ECUs poured into the quest for the ultimate particles of nature.

Not only has atomism become triumphant, but the other aspects of the scientific revolution have also flourished. Modern work has seen the progressive extension of Descartes' programme for explaining many human actions in mechanical terms. Now even the subtle actions that Descartes attributed to the soul are accounted for in terms of the brain, using computer analogies to try to understand how reasoning and speech could arise by purely mechanical means. If that could be achieved, then Descartes' soul would be rendered completely redundant and the universe would become a pure geometrical machine.

This history has, however, been interrupted by the second, modern, revolution in our concept of reality which I shall describe in Chapter 4. First, however, we need to look more closely at that second plank of atomism, the void, and its relation to space and time.

Chapter 3
Space and Time

1. The void in Aristotle

Early in the seventeenth century, Evangelista Torricelli investigated the mercury barometer: a vertical tube containing mercury, closed at the top with its bottom in a vessel of mercury. The column of mercury partially fills the tube and above it there is a vacuum, so that the weight of the column is supported by the pressure of the atmosphere on the surface of the mercury in the vessel. This 'discovery' of the vacuum has often been cited as a crucial experimental disproof of Aristotle's physics, since Aristotle, it is said, claimed that there could be no vacuum in nature. It is thus seen as an integral part of the revolutionary overthrow of Aristotelianism. As with the introduction of atomism, however, the history of thought about the void turns out to be much more subtle than the simple notion of accumulating experimental evidence would suggest.

Aristotle's view of the world, it will be recalled, was based on a careful logical analysis of our actual experiences. He avoided, on the whole, speculative theories and he avoided the Platonic approach of giving absolute reality to ideas and to mathematical entities. For him the universe consisted of a collection of things involved in processes of change.

His word for void literally means 'the empty,' the 'no-thing,' and it is this that he claims is impossible. Because it is a concept defined within Aristotle's own system, we cannot assume that it means the same thing as 'the vacuum' of modern physicists or 'the void' of the atomists. It is worth looking closely at what it actually does mean, for that will shed a lot of light on the whole notion of space that is essential for atomism.

Aristotle's discussion of the void occupies Chapters 6-9 in Book IV of *The Physics*. One difficulty is that it appears to be very unbalanced. He spends almost all his time examining other thinkers' arguments in favour of the void, and rejecting them, showing that one can make sense of the universe without believing in the existence of the void. So this ought to lead him to the position that the void is an optional extra, something that might exist, but which is not necessary. In fact, however, he clearly takes the view that the void is impossible.

This could be just a matter of economy, of not believing in anything unless there is a positive reason for doing so. But a careful reading of his arguments shows that there is more to it than this: central to the whole discussion is the fact that his basic physical approach makes it impossible for there to be a void in the sense of 'the empty':

> To determine whether the void exists or not, we must know what they who use the word really mean by it. The current answer is, *a place in which there is nothing.* (213b 30)

So in order to decide whether such a thing can exist, we need to know what is meant by 'a place' and what is meant by 'nothing.' Aristotle continues by enlarging first on 'nothing':

> But that is in the mouths of people who consider that
> nothing 'exists' except material substances;... Again they
> regard all corporeal substances as 'tangible'; and this they
> regard as identical with 'possessed of gravity or levity.'
> Thus we may logically expand their definition of the word
> [void] as 'that in which there is nothing that is either heavy
> or buoyant.'

(The reference to heaviness and buoyancy belongs to an obsolete system of thought, but the essential idea here is that of tangibility which is still very much part of our common way of thinking.) It then becomes clear that Aristotle has no problem with this reduced concept of void; he will shortly come to the more absolute concept that he claims is impossible.

Now this reduced concept of the void is essentially the scientific concept of the vacuum — the gap at the top of Torricelli's barometer tube. It is a place where there is nothing 'tangible'; but it is not a place where there is nothing at all. On the contrary, a vacuum is permeated by electric fields and gravitational fields, and in the modern concept of the void is often regarded as a highly structured physical entity that contains in it the potential information for all the fields that can traverse it. It is very far from nothing.

What is impossible, Aristotle then continues, is a place that contains *absolutely nothing.* And this is because of the particular way that he regards 'place.' For him, a place is not part of an abstract mathematical space, because he has rejected absolute mathematical entities of the sort espoused by Plato. Instead, a place is always the place of something: it is the geometrical shape delineated by the boundary of a particular thing, by the surface that forms the interface between the thing and the other bodies (air or whatever) that surround it. Where there is no thing, there is no place. So the idea of a place without a thing is self-contradictory. Aristotle puts it as follows, using the term 'body' for 'thing' (where, in view of what has gone before, by 'body' he does not necessarily mean something tangible:

> ... since the void (if there is such a thing) must be
> conceived as place in which there might be body but is not,
> it is clear that, so conceived, the void cannot exist at all ...
> for what is meant by it is not body but the bodiless
> dimensions of body. (214a 20)

To make clearer what Aristotle is saying, let us apply this idea to the vacuum at the top of the barometer. The Greeks knew nothing about fields, so in their place we might postulate that this vacuum really was an absolute void, total nothingness. Then, Aristotle would respond, what is it that separates one side of the barometer tube from the other? If there is nothing in between, then the two sides of the tube would be the same surface and the tube would in fact be a solid rod with no internal space. As soon as we say that the tube has an internal surface — a boundary between the glass of the wall and whatever it is that constitutes the internal space of the tube — then we are postulating what Aristotle would call a 'body,' a something, between the two sides of the tube.

In other words, what Aristotle is doing is not denying the existence of a vacuum such as exists at the top of a barometer; but he is pointing out that this vacuum, even though it is intangible, is actually a very specific physical entity that should not be taken for granted. This is a profound physical insight with implications which are only now starting to be grasped by physicists.

Where, then does this leave the atomists, both Greek and seventeenth century? It means that one cannot dismiss the void as being 'just empty space.' In fact 'space' is simply another word for 'the void' as used by atomists. Whether called space or void, Aristotle's argument points to the fact that it has to be treated as an entity with physical properties, and not swept under the carpet. When the atomists say that the universe consists of atoms and the void, they are not saying in poetic language that it consists of particles and, apart from that, nothing: the void (space) is just as much a physical entity as are the particles, and both together are needed for atomism.

2. The void and relations

Aristotle's analysis had no impact on the development of physics at all. In retrospect this was not surprising: to handle such a sophisticated concept of 'place' quantitatively requires, even today, every resource available to modern mathematics. The questions that he raised were debated quite acutely by Islamic philosophers (see Sorabji 1983); but in the West they largely lay dormant until they were raised again by Leibniz and Newton in the debate that I will analyse shortly. The terms of reference had by this stage changed, and it was the new agenda set by this debate that subsequently formed the subject of analyses by Kant in the eighteenth century, and by a host of modern philosophers (Grünbaum, Lucas, Swinburne and others).

The importance of all this for us is that, despite these many searching analyses, the atomistic concept of the void (space) has remained an unshaken component of the scientific concept of reality.

Space is the most basic aspect of this modern concept of reality, and yet it is the aspect that has perhaps the weakest philosophical foundation.

Before looking at the seventeenth century debate, it will be helpful to get a clearer picture of the issues, and the way in which Leibniz broadened them. For this, we need to go back to the idea of 'place.' For Aristotle this was the bounding surface of a body; but in popular use something different seems to be implied.

Image a child who brings to you a vase that she has picked up from a table, and you tell her to put it back in the same place from which she got it. By 'the same place' you mean 'on the table,' and with luck the child might return it there. But if she had recently been fired by an enthusiasm for astronomy, or by a desire to irritate her parents, she might argue that in the four minutes since she picked up the vase the earth would have rotated by one degree, so that the place from which she picked it up would now be occupied by a point in the Atlantic ocean and inaccessible to her. Indeed, she might also take into account the fact that the earth would have moved in its orbit round the sun by some 7000 km; and indeed that the sun would have moved in its orbit in the galaxy by a greater amount. One could go on to include the motion of the Galaxy in the local cluster of galaxies .... Where would the process stop?

What this shows is that 'the same place' does not have a fixed meaning. It means one thing at home, and another thing — in fact several possible other things — for an astronomer at work. If you want to be precise, you have to specify what the place is *relative* to. Your original instruction actually referred to the same place relative to the table (or relative to the house), not relative to the centre of the earth, or to the sun, etc. In cases like this 'place' is a relative concept. So, if by 'space' we mean the collection of all possible places, then space is also a relative concept.

This discussion brings in a new concept that was not explicitly recognized by Aristotle: the idea of place as a relation. A place, on this analysis, is always a place relative to the sun, the earth, the table or whatever, and so it is to do with relations. Aristotle focused just on the relation of contact, defining place in terms of the boundary where one body touched another. But relations could also be things like 'between' or 'behind'; and they could be extended to include numerical components, such as 'two metres in front of' — though such an extension then raises all sorts of subsidiary questions.

3. The views of Leibniz and Samuel Clarke

Here we come to focus on a famous debate about time and space that took place in the seventeenth century between Gottfried Leibniz and Samuel Clarke. Clarke (no relation) propounded Newtonian views and may even have consulted Newton at times; Leibniz was an independent minded philosopher, who had broken free of earlier scholasticism but was not prepared to swallow uncritically the system that Newton had developed. His criticism hints at a route that science might have taken, if the forces of history had been different.

Leibniz accepted the argument outlined above as basic, holding that 'space' always meant 'relative space.' Newton, however, believed that when one followed down the chain of 'relative to the earth,' 'relative to the sun,' and so on (he did not know about galaxies), one could eventually reach some ultimate meaning of 'in the same place' that was absolute, and not relative to anything. This absolute space was the equivalent for him of the void of the atomists. He formulated this belief, and the distinction between absolute and relative space, in the 'Scholium to Definition VIII' of his *Principia Mathematica Philosophiae Naturalis,* a statement that lies behind the whole debate. It is worth reflecting that this work is the foundation for the mechanistic science that has dominated thinking ever since — a blockbuster of a book, deliberately designed to carry as much authority as possible; written in Latin, at a time when this was going out of fashion, in the formal style of Greek masters such as Euclid, propounding not theories or opinions but Laws decreed by the inexorable rule of the Almighty Father. In translation, Newton wrote:

I do not define time, space, place and motion, as being well known to all. Only I must observe, that the common people conceive those quantities under no other notions but from the relation they bear to sensible objects. And thence arise certain prejudices, for the removing of which it will be convenient to distinguish them into absolute and relative, true and apparent, mathematical and common ...

Absolute space, in its own nature, without relation to anything external, remains always similar and immovable. Relative space is some movable dimension of measure of absolute space; which our senses determine by its position to bodies; and which is vulgarly taken for immovable space;... For if the earth, for instance, moves, a space of our air, which relatively and in respect of the earth remains always the same, will at one time be part of the absolute space into which the air passes; at another time it will be another part of the same, and so, absolutely understood, it will be continually changed ...

Absolute motion is the translation of a body from one absolute place into another; and relative motion, the translation from one relative place into another. Thus in a ship under sail, the relative place of the body is that part of the ship which the body possesses ... and which therefore moves together with the ship: and relative rest is the continuance of the body in the same part of the ship But real, absolute rest, is the continuance of the body in the same part of that immovable space, in which the ship itself, ... and all that it contains, is moved. (in Alexander 1956, p.152f)

Newton is here expounding his version of 'the void' of the atomists, extending infinitely in all directions. He uses the key terms 'absolute' and 'mathematical.' Space is absolute as opposed to 'relative' or derived. Newton's space is quite independent of what is in it; the universe could be completely without matter and space would still be there. He seems to have regarded it as a kind of property of God (this forms the subject of part of the correspondence between Leibniz and Clarke) though he expresses himself very cautiously on this point, rightly realizing that he is in danger of becoming dangerously heretical in his theology.

Space is also mathematical, a Platonic idea echoing Descartes' belief that ultimate reality is mathematical and geometrical (although Descartes was, as we have seen, not a strict atomist). Thus space was — as Aristotle's argument requires — a real thing that expressed the laws of geometry taught by the Greeks and developed to new heights by Newton himself. Human beings might make physical measurements that gave approximations to the real laws of geometry, but these laws were actually enshrined in space itself, independently of our attempts to measure them.

Newton goes on to give various arguments for believing in such a space, none of which seems very convincing, to which we shall come in a moment. It is this concept of space that comes to form one of the central themes in the extensive exchange of letters between Leibniz and Clarke. The discussion beautifully illustrates the central debate that has gone on ever since between those who have quite opposite conceptions of reality. Leibniz regards as real only things that are based directly on our perceptions; Clarke (following Newton) regards as real the mathematical 'absolutes' that lie behind our senses. Leibniz thinks that this mathematical absolute space is nothing but a theoretical construction, something in the imagination of mathematicians; Clarke thinks that Leibniz' reliance on the physical bodies perceived with our senses involves mistaking appearance (what we sense) for reality (what we can know intellectually). The whole exchange was published by Clarke, with Leibniz' letters (originally in French) carefully translated by Clarke.

4. The debate

Initially Leibniz bases his argument on 'the principle of sufficient reason,' which says that nothing happens without a sufficient cause. It is very much a concept of the enlightenment, based on the conviction that the universe must be rational. Everything must have a meaning, and God does not play dice. Clarke, too, accepts this principle and so they have a common ground for their argument. We take up the debate in Leibniz' third letter, after he has declared his opposition to the idea of 'real absolute space' which, he says, 'is an idol of some modern Englishmen.' He, by contrast, holds that space is not a 'thing' at all, but just a shorthand way of referring to the positions of objects relative to one another (what Newton calls relative space). He writes:

> If space was an absolute being, there would something happen for which it would be impossible there should be a sufficient reason. Which is against [the principle of sufficient reason]. And I prove it thus. Space is something absolutely uniform; and, without the things placed in it, one point of space does not absolutely differ in any respect whatsoever from another point of space. Now from hence follows, (supposing space to be something in itself, besides the order of bodies among themselves,) that 'tis impossible there should be a reason, why God, preserving the same situation of bodies among themselves, would have placed them in space after one certain particular manner, and not otherwise; why everything was not placed the quite contrary way, for instance, by changing East into West. But if space is nothing else, but that order or relation; and is nothing at all without bodies, but the possibility of placing them; then those two states, the one such as it now is, the other supposed to be the quite contrary way, would not at all differ from one another. Their difference therefore is only to be found in our chimerical supposition of the reality of space in itself. But in truth the one would exactly be the same thing as the other, they being absolutely indiscernible; and consequently there is no room to enquire after a reason of the preference of the one to the other. (Alexander 1956, p.26)

Leibniz here starts off with a theological argument involving the principle of sufficient reason. If there is such a thing as absolute space, and given a common belief in God as the creator to the universe, then there needs to be a sufficient reason for God to have created the universe in one position, rather than in another position, or with a reflected arrangement of everything. He then moves on from the theological argument to what is actually the core of his position: the argument that it is absurd to think that two arrangements of the universe which are completely indistinguishable should in some mysterious way be different. This position is based on the principle that reality has to be based on things that can actually be checked out somehow, so that if two things cannot be in any way distinguished, they must be the same thing.

The theological argument falls flat. Clarke simply replies that there is a sufficient reason for the universe being in one place rather than another, namely the will of God:

> The uniformity of space does indeed prove, that there could be no (external) reason why God should create things in one place rather than another: but does that hinder his own will, from being in itself sufficient reason of acting in any place, when all places are indifferent or alike, and there be good reason to act in some place? (p.32)

Both parties pursue the question of what the reasons of God might be, but, since neither has a hotline to God, this particular argument is destined to get nowhere. The second of Leibniz' arguments, based on the idea that the only real things are those that actually make a difference to the way things are, a difference that can be checked with our ordinary senses, is more fruitful. It prompts Clarke either to look up the passage in the *Principia* that I quoted earlier, or actually to talk to Newton. As a result, Clarke takes the argument into the area of physics. He implicitly recognizes the principle that if space is real then it must make an observable difference. But this difference, he claims, is not to be found by examining the (static) relations of bodies among themselves, but by looking at the dynamics of bodies — the way they move (a strand of argument that had been raised earlier, but only in a theological context similar to the argument we have just been examining). He now writes:

> Two places, though exactly alike, are not the same place. Nor is the motion or rest of the universe the same state; any

more than the motion or rest of a ship, is the same state, because a man shut up in a cabin cannot perceive whether the ship sails or not, so long as it moves uniformly. The motion of the ship, though the man perceive it not, is a real different state, and has real different effects; and, upon a sudden stop, it would have other real effects; and so likewise would an indiscernible motion of the universe. (p.48)

This, from a modern perspective, seems to touch on the key issue. As Newton had explained, if one was shut up in a ship then everything in the cabin is *relatively* at rest, relative, that is, to the ship. But we can ask, is the ship and everything in it absolutely in motion or not? Leibniz would say the question has no meaning, since only relative motion is something real; Clarke is saying that one can ask the question, and answer it: if it is absolutely in motion, then, if the ship were suddenly to stop, everything would be thrown forwards in the cabin; but this would not happen with merely relative motion.

Leibniz' defence to this reasoning is unconvincing, partly because he has not, as we have, had the benefit of three hundred years thinking out the implications of the Newtonian physics that had just burst onto the intellectual stage. First, he reiterates his philosophical position: that if the *entire* universe were to 'move,' then this would be a motion in which nothing changed, and so must be equivalent to rest; hence, whatever Newton might say, it is impossible that there should be any detectable difference, either directly or through looking at dynamics, between such a motion and rest. This is all well and good, but it does not explain where Clarke's argument might be wrong.

Second, he has a line of argument that retreats to a position that can be found in Aristotle, in making a difference between (relative) motions where the cause of motion is in the body, and where the cause of (relative) motion is elsewhere. This gets us some way towards understanding Clarke's distinction between a boat in motion and a boat at rest. If the boat hits a rock, when we would say that it 'suddenly stops,' the cause of the event is the contact with the rock, which is clearly sticking into the ship. What occurs is a change in the relative motion of the ship and the things in it (they are thrown forwards), but the cause of this change is the rock, which is touching the ship, not the things in it, and so the appropriate explanation is to be sought in a change in the motion of the ship, not its contents. There is no need to appeal to an abstract idea like space when one has a solid rock to hand. Though this is hopeful, it still has a long way to go before it starts being competitive with Newton's approach: it is all very well using qualitative ideas like 'the cause of motion being in the body,' but how does one get a quantitative physics out of it? And how will the idea fare when one gets into more complicated cases, in which, for example, all the objects concerned are being acted on by different sorts of force-fields?

From a modern perspective and using modern terminology, Clarke's argument is roughly as follows: When a ship stops, the deceleration has a clear physical effect. The same is true of acceleration, when everything is thrown backwards. So (absolute) acceleration and deceleration are certainly real. But acceleration is change of motion, and so if absolute acceleration is real, so is absolute motion. Further, motion is change of position, so if absolute motion is real, so is absolute position. Put like this, we can see exactly how the argument is only half correct. Certainly, we have absolute acceleration. We can then look for the 'absolute motion' whose change gives the observed absolute acceleration; but *there will be many 'absolute' motions corresponding to a given acceleration.* If I am in a ship's cabin and I experience myself and everything in it thrown forwards, it might be because we had been moving forwards and suddenly stopped, or it might have been because we had been stationary and suddenly started moving backwards, or any intermediate possibility. We have a whole range of 'absolute motions' that could be deduced from a given absolute acceleration, which makes the term 'absolute' rather nonsensical when applied to the motions, and certainly makes it impossible to carry out the next stage of the argument, which requires there to be a single absolute motion from which to deduce absolute space.

5. The verdict of history

If my tracing of the arguments is correct, then philosophically it looks as though Clarke's argument does not stand up, while Leibniz' objection to the idea of absolute space seems sound. Does this judgment follow through from philosophy into science, however? Newton has presented an elegant scientific theory that relies on the idea of absolute space, involving laws that say that bodies continue in a state of uniform absolute motion unless acted on by forces. Can Leibniz produce as good a theory without using absolute space? No, he couldn't. So does this mean that, scientifically speaking, absolute space is real? This brings us to the key question underlying the whole of the modern predicament about reality. Does the success of a scientific theory imply the reality of everything that is in that theory? Answering 'Yes' to this is the approach of *scientific realism,* which holds that the true touchstone of reality is whether or not a given concept 'works' in some scientific theory. The scientific realist would have to answer, yes, space is real, not because Clarke has won the philosophical argument, but because Newton can deliver the scientific goods and Leibniz cannot.

This now starts to become worrying. Reality is being decided on the basis of a sort of trial by combat, in which the judgment goes to the victor. What is worse, the rules for the combat seem arbitrary: if Leibniz has won the philosophical argument, and Clarke the scientific argument, why should we say that the scientific argument 'counts' and the philosophical one doesn't? And what would have happened if Leibniz had been a better physicist, and had come up with a better theory than Newton without using absolute space?

In a way, we can answer the last question by bringing Einstein into the combat. His special theory of relativity in fact does just this. It does not use the same principles as Leibniz (how to turn Leibniz' arguments into scientific theory is a different, and very fascinating, story), but it does base itself on relative rather than absolute motion. Indeed, the whole theory is based on the fundamental principle that Newton is wrong: it is taken as an axiom that there must be nothing in physics that allows one to define an absolute space! It is this explicit denial of the main ingredient of Newton's theory that leads, together with facts about electromagnetism unknown to Newton, to a theory of dynamics which turns out to be much more accurate than that of Newton. If the scientific realist concluded in 1700 that absolute space was real, then in 1905, a realist would have to conclude that absolute space was not real. Can such a basic part of reality fluctuate in this way? We have here a perfect illustration of the questions already examined of the relationship between the history of science, with all the changes in scientific world-view that have taken place, and attempts to pin reality on scientific theory (see Chapter 1).

The history of the idea of space suggests that it is dangerous simply to assume that all the ingredients of a scientific theory are real just because the theory 'works.' But there are more specific lessons to be drawn from this piece of history, which relate directly to modern physics with its stress on space-time, as opposed to space.

The basis of relativity theory is to stop working with (absolute) space and (absolute) time, but to work instead with space-time. Just as space was the extent of all possible places, so space-time is the extent of all possible events. Any given observer (that mythical inhabitant of science textbooks, whom we need to interrogate more closely later) will divide space-time up into space (position relative to her/him) and time (also relative). But different observers will do it in quite different ways. So far this seems all quite reasonable, even though it is impossible to see from this bald description what one might get out of such a position. The next step, however, is far from reasonable and is usually taken very covertly. Physicists had become so used to working with Newton's absolute space, that they could not conceive of operating without having something as an absolute: a given substance, fixed and laid down by God, which provided a container for everything that happens in nature, described by mathematical laws. So when Einstein abolished absolute space, it was immediately replaced by *absolute* space-time.

Instead of realizing that science had been misled by the apparent success of Newton's theory into accepting Newton's metaphysics, physicists simply transferred this same metaphysics to Einstein's space-time. Instead of realizing that the way was now open for taking Leibniz seriously, and giving due weight to his philosophical argument, they continued to adopt scientific realism, shifting allegiance to absolute space-time, like a herd of deer when a new stag conquers the old. As a result we now have a host of attempted scientific theories in which space-time is regarded as a thing, like a table-cloth stretched out across the universe, capable of waving, bending and even coiling itself up into tunnels.

The inheritance of this absolute notion of space-time has a double significance. First, as I noted earlier, it underlies our whole notion of reality. Consequently, we have no alternative framework in which to fit experiences that might suggest an alternative reality. Telepathy or clairvoyance, even if they could be established in terms of evidence, cannot be handled by science because it has to think of them in terms of messages propagating across Newton's absolute space. Such a pattern of thought is clearly alien to the nature of telepathy which, judging from the anecdotal evidence, involves a sort of rapport between people in which a relation of closeness is established that does not fit into Newton's absolute space. Similar problems hold for the corresponding phenomena in the domain of time, particularly precognition, where again we seem to be dealing with a relational structure that cuts across Newton's mathematical time. Reality, in the sense of what the world is like, could well need to incorporate such experiences.

The second significance of the Newtonian inheritance concerns the development of physics itself. The greatest unsolved problem within the agenda of conventional physics is that of combining Einstein's theory of general relativity with quantum mechanics (the latter will be examined in the next chapter). Increasingly many mainstream physicists are now arguing that this can only be done if the picture that we have of space and time is altered. There have been many attempts to modify the model used for space and time, some of them very radical: the number of dimensions has been altered, the number system on which the mathematics of space is built has been modified, and so on. But it is now starting to look as though the idea of a continuous mathematical space, existing in its own right as something in addition to matter, is the basic obstacle to finding a theory that produces finite answers.

General relativity proved enormously successful in taking scientific thought beyond Newton and opening up a vast new area to our understanding. But the metaphysics of Newton now acts like a straightjacket, forcing us to fit everything to a four-dimensional version of Newton's void and making it impossible for relativity theory to grow and adapt to the challenges posed by quantum theory. It is ironical that the very name 'relativity,' associated with the removal of Newton's absolute space, is attached to a theory that is insistent that space-time is absolute, affected by the matter in it (unlike Newton's space) but having an existence that is essentially independent of matter.

While the original void of the atomists was a bold and coherent concept, its successive transformations into the absolute mathematical space of Newton and the dynamical absolute space-time of Einstein have given rise to a metaphysics that now seems incompatible with quantum mechanics. For decades theoretical physicists have been struggling to find a mathematical fix to escape from the impasse into which they have wandered. The time has come to rethink the whole course of history that has led them there.

Chapter 4
Quantum Theory

In Chapter 2, I described atomism, the doctrine of atoms and the void, the two planks on which science has rested. We have seen how the historical origins of atomism make it impossible to hold the view that atoms and the void are simply reflections of an absolute reality. The previous chapter showed that the idea of the void, as it has come to be expressed in terms of space and time, is very uncertain. Now we shall examine how quantum theory demolishes the other plank of atomism, the idea that there really are concrete atoms with well-defined positions and speeds constituting reality 'out there.' At the same time, quantum theory opens up completely new ways of thinking about the universe.

For the last three centuries the Western world, and more recently the entire developed world, has been dominated by physical science; and physical science has been dominated by the Newtonian vision of the universe as a vast machine working until eternity according to fixed and inexorable laws. Reality was a machine. In this century, however, a new idea has emerged: quantum theory. Despite a flood of books on the subject, its implications are barely beginning to be understood, least of all by the experts who apply it in physics. As a result of this idea, our view of reality will never be the same again.

So, what is quantum theory? The standard answer, to be found in textbooks of physics, is that it is a theory used to describe very small particles of matter — atoms, electrons, protons and the like. It began with the attempts of physicists to understand the operation of atoms (the atoms, that is, of modern chemical theory, not the ultimate particles of Lucretius and Democritus). It was realized that atoms were made up of a tiny nucleus around which moved a number of charged particles called electrons. But a moving charged particle, according to conventional physics, should emit all its energy in the form of light and rapidly come to rest; so how was it that atoms survived in a stable state? And why was it that, when they did emit light, they only emitted light of one of a precisely specified number of colours?

These were the sort of questions that provoked physicists to invent a very new sort of physics: quantum theory. If quantum theory were confined to the details of atoms, then there would be no need for non-physicists to worry about the subject. There was, however, nothing in the theory that said that it could not apply to other areas as well. Indeed, physicists have for some time been investigating large-scale objects that can only be understood by using quantum theory: super-conducting electromagnets in which an electric current flows round and round for ever without dying away (just as the electron for ever circles the nucleus); or super-fluid liquids that can creep through the tiniest crack without any resistance.

Even before such examples emerged, a few courageous writers such as David Bohm grasped and expounded an alternative view of quantum theory. It is not just a branch of physics adapted to a few exotic sorts of phenomena, but is a more general way of talking about reality than we have had before this century, a new language and a new way of thinking, forced on us by the physics of small particles, but applicable everywhere. Perhaps in fifty years' time, children will be learning quantum mechanical thinking through nursery stories. Today, however, the ideas are still too new, too foreign to old ways of thinking, for us to see fully how they link in with life. So we still have to go back to the origins of quantum mechanics in physics and struggle to extract from the physics the implications for humanity. Perhaps we need a new Lucretius who can paint in dramatic literature the new world that is opened up by quantum theory.

The theory began in a conflict between two different ways of thinking about the physical world: as made up of particles, and as made up of waves. Neither of these seems very meaningful to most people, for whom the world is made up of humans, tables, governments and so on; but it is the physicist's self-appointed task to examine the world under a microscope, shutting out these larger meanings and focusing on the smallest constituents of things. Sometimes these seem to be particles, and sometimes waves. Quantum theory reconciles the two through a third alternative — an abstract, unvisualizable thing called a quantum state. The world is not 'really' waves and it is not 'really' particles. Is it really a quantum state?

In this chapter I shall introduce the players in this conflict — waves and particles — and then describe their roles in quantum theory.

1. Waves

Waves on a pond create larger patterns as they ripple through each other, in some places building up to a peak of activity, in other places cancelling each other out to leave quiet water. This interaction is called interference. In its simplest form, two vibrating objects (such as the two prongs of a tuning fork) dipped into a pond will each create circular spreading waves that interfere to form a series of radiating stripes of activity and inactivity (Figure 1).

From the early days of science it was known that light, passing through narrow slits or reflecting from surfaces very close together, could show similar patterns. A typical experiment is the two-slit experiment for light, in which one shines light of a pure colour through two slits cut parallel and close together in a metal plate. When the light falls on a distant screen, it displays a series of closely spaced parallel stripes, reminiscent of the stripes of activity radiating from the prongs of the tuning-fork in the pond. So waves could explain the way light behaves; and in the nineteenth century this was made the basis of a detailed theory in which a full account of light was worked out and tested in terms of waves of electricity and magnetism (electromagnetic waves).

2. Particles

We have already examined (see Chapter 3) the way in which the seventeenth century scientists were attracted to Greek ideas that matter was made up of tiny indivisible particles called atoms moving through an infinite void. For whatever reason, the belief emerged that the world could be understood entirely in terms of the way atoms interact

Figure 1. The interference patterns produced by two nearby sources of waves.

with each other, and that these interactions obeyed simple mathematical laws. In one of the most dramatic leaps of faith in cultural
history, it was proclaimed that humanity could, by mathematical
reasoning, understand the very foundations of the universe. It was a
heroic stand, in the mould of Prometheus who stole the fire of the
gods.

The Promethean programme has held sway unchanged ever since,
but the terminology has changed in a confusing way. For the Greek
atomists, 'atoms' were the fundamental indivisible particles that held
the ultimate explanation of everything. We now call these 'fundamental particles' and we use the word 'atom' to mean the smallest
part of a chemical element. Our 'atoms' are divisible, being made up

of protons, neutrons and electrons. The basic philosophy, however, is unchanged: the ultimate explanation is to be found in the interactions of the smallest constituents of matter. For simplicity, I shall continue to call this philosophy 'atomism.'

3. The clash of paradigms

Quantum theory was born when, in a series of critical experiments, the two paradigms of wave and particle came into conflict. First, in 1916 Millikan measured the energy of electrons given off by a metal when light shone on it. It was a tricky experiment. The only metals that behaved in this way were very reactive metals that tarnished immediately in the air, and so in order to get reliable results the metal had to be cut by remote control in a vacuum in order to produce an absolutely clean surface. The electrons given off were detected electrically, and their energy was measured by finding the strength of the electric field that was needed in order to repel them from the detecting instrument.

The phenomenon was not in itself very mysterious. As we have seen, light was explained as a wave of electricity and magnetism, and so it would be capable of attracting the electrons in the metal and drawing them out. On this way of thinking, the stronger the light, the more powerful the electric effect, the more energetic the electrons. What was found was at first completely inexplicable: the energy of the electrons remained unchanged as the intensity of the light increased, only the total number of electrons increasing. Their energy turned out to depend, not on the intensity of the light but on its colour.

Eventually, by performing many other experiments in which light interacted with particles, a solution to the problem emerged. Although light seemed, in the two-slit experiment, so clearly a wave, when it expelled electrons from a metal it behaved like a stream of particles. Each particle of light (called a photon) had a definite energy that depended on the colour of the light, and each photon knocked one particle out of the metal. So the energy depended on the colour, and the number on the intensity. It was neat, but completely contradicted everything that people had previously thought about light.

Once light was seen to behave as particles, the stage was then set for the reverse effect: particles behaving as waves. In 1929 Davisson and Germer observed that electrons scattered from crystals in a wave-like manner. There were alternating stripes of intensity emerging from the crystal, where the regular arrangement of atoms was acting like a series of slits for an interference effect. Later it proved possible to perform the two-slit experiment with particles (it was first performed with electrically charged atoms), just as with light. So light and electrons could each behave either like waves or particles, depending on the circumstances.

4. Quantum language

At this point physics was faced with a breakdown in its conventional methods of description. There was a period of anarchy. Writers were at first aghast, then decided to make a virtue of illogicality, stressing in popular books the paradoxical nature of what seemed to be happening. Some writers felt that the breakdown of conventional descriptions marked the end of rational physics, the attainment of the edge of the knowable.

Many still speak in this way today; but, gradually, there came a realization that the problem was not an absolute breakdown of rationality, but a breakdown in the particular ways in which physics, and, following physicists, everyone else, was choosing to talk about the world. A new language was needed, with new concepts, to be provided by quantum theory.

Current physics associates with every physical thing a *(quantum) state*. A state is neither a wave nor a particle. It is neither observable by physical experiments nor visualizable in pictorial terms. Such phrases start to make the state look like a deep and mystical concept, like the Tao; but this is far from the case. The quantum state, though it is unfamiliar, is also very concrete. We can deduce lots about it indirectly from the observable effects that flow from it. Whenever we make a measurement we find out something about the current state of the thing measured, although we change that state in the process. Most importantly, there is a mathematics of states: states can be combined together in variable proportions so as to define new states called superpositions of the old ones. The emergence of an actual observation from the quantum state does not depend on the state alone. It depends on us as well. We enter a dialogue with the state when we set up an observation, and the outcome is a joint production arising from this dialogue. The questions we ask determine, in part, the outcome. Thus, by placing slits in the path of the light, we are asking a question that requires an answer in terms of waves; by placing electrons in the path we ask a particle-question. But the underlying state is neither wave nor particle.

David Bohm, referred to above, used the phrases 'implicate order' and 'explicate order' to describe this situation, though his approach differs in several crucial respects from the conventional one. The implicate order is the realm corresponding to the quantum state which contains a wealth of potentiality folded up (implicate) inside it, only some of which emerges (explicate) to the observed world. I shall explore his penetrating analysis later (see p.102).

At this stage many people suspect a confidence trick. We seem to be lured into a position where we are postulating a mysterious shadowy realm of the implicate order, neither real nor unreal, and abandoning our well-tried ideas of a solid physical reality underlying all we see. We would be right to think very carefully before we go down such a path. Could it be that we do not have to revise our concepts of reality at all?

There is an alternative position (favoured by Popper, for example), according to which the 'quantum state' is something quite familiar: it is the sum total of our knowledge of some ordinary physical state-of-affairs which happens to be hidden from us because of the limitations of our measurements (but knowledge that is wrapped up in a mathematical form). An example will make this clearer. If I throw a dice (or, for statisticians and pedants, a die) and hide it in my hand, then your knowledge of the top face is initially only that it is a number between 1 and 6, inclusive. Suppose now I give you a glimpse of it, from a distance, enabling you to make out that there are at least four spots on the top face. Your knowledge has now changed as a result of the observation; you can now say 'the number is 4, 5 or 6.' What was previously pure potentiality (any number from 1 to 6) has become more concrete (4 to 6). If I were to tell you now that the number was even, your knowledge would change again, to knowing that the number was 4 or 6.

We could describe all this in quantum mechanical language. If by 'state' we mean 'knowledge,' then the state changes when we make an observation. The nature of the change depends on the observation we make (whether a distant glimpse or a clue from me). And at the end we find the state in a superposition between the state of being a 4 and the state of being a 6. But using this language does not alter the fact that all the time there has been a well defined real world in which the number has been 6 all the time. Why should not exactly the same thing be the case for photons and electrons?

There are many arguments against this, of which I shall give the two most important, concerned with *complementarity* and *non-locality*. Both involve further experiments, which the reader who is mainly interested in the philosophical conclusions may wish to skip over.

5. Complementarity

The crucial difference between knowledge and the quantum state is that knowledge of a traditional reality is cumulative. The traditional image of science is that, by piling up more and more knowledge about some state-of-affairs, one approaches ever more closely to a complete specification of what is going on. We might refine this picture by admitting that there may be some things which, because of essential physical limitations, we can never know; but it will always be the case that knowledge will get us ever closer to whatever this limit is. One piece of knowledge can never contradict a previous piece, unless one of them is simply wrong.

It is quite different with observations of a quantum state. Here the different observations we can make (the different questions in the dialogue, to use our earlier language) lead us along different mutually exclusive paths. It is somewhat as if the universe is like one of those ambiguous pictures, such as that shown in Figure 2, which can be interpreted either as a duck or a rabbit. Our brain must choose bet-

Figure 2. An ambiguous picture.

ween one of the two interpretations, each of which rules out the other, with no middle ground.

Similarly, when a physicist is carrying out an experiment in which quantum theory is dominant, a choice must be made between mutually exclusive aspects that can be explored. One piece of 'knowledge' rules out another piece that we did not choose to explore. So two (or more) exclusive paths are not passively collecting information about something existing as an independent reality, but they are themselves creating different realities that were both potential (implicate) in the original state. Paths of this sort are called complementary. The wave and the particle paths of investigation are complementary in this sense.

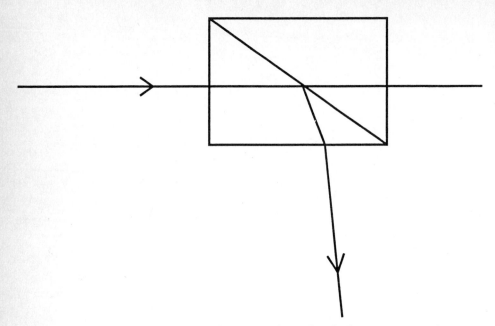

Figure 3. Photons passing through a prism that splits the beam into two parts.

Physicists have developed a wide range of experiments, in which the arrangement of the experiment selects one particular aspect of a physical system, thereby ruling out other aspects. It is useful to compare two of these, one where the physical systems are photons (particles of light) and the other where they are electrons.

In the photon experiment, the light is directed onto a prism with specially coated faces (Figure 3) that splits the light into two beams in which the electrical vibrations that make up the light (on the wave picture of thinking about it) take place in two different directions. We could, for instance, arrange things so that in one beam the electrical vibrations were taking place horizontally, while in the other they were taking place vertically. If we were to rotate the prism, then these two directions would also rotate, while keeping the same relation to each other and to the prism. If, now, the intensity of the beam of light is reduced to the point where it consists of individual photons passing through the prism one at a time, then each photon will emerge either in the horizontally-vibrating beam or in the vertically-vibrating beam. Since photons are indivisible, there is no alternative. But if we had set up the prism at a different angle, then the division would not have been into horizontal and vertical, but into two different angles. There are infinitely many different angles available to us in which to set up the apparatus, and each represents a different path of exploration. The paths are complementary, in that our choice of one angle rules out every other angle.

The other experiment is called the Stern-Gerlach experiment, and is similar in that the different paths of investigation correspond to different angles of the apparatus (Figure 4). If electrons are directed between the poles of a magnet that is set up to produce a non-uniform field, then the electrons are deflected. They behave like small bar-magnets themselves — indeed the magnetism in a bar magnet is due to the combined effects of many electrons. We can suppose the field is arranged to point vertically and to increase in strength vertically, in which case it turns out that the electrons will move up or down. If we were simply to rotate the magnet producing the field so that it was horizontal, then the electrons would move to the left or to the right. None of the electrons go straight on; all are deflected, one way or the other, by the same amount.

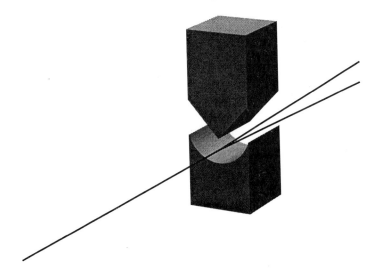

Figure 4. The Stern-Gerlach experiment. Electrons enter from the left and pass between the poles A and B of a strong magnet, being deflected either up or down.

Having described these experiments that set up one out of many complementary paths, I can now explain the difference between the old picture, of a fixed reality with cumulative knowledge, and the new picture with alternative options selected by participation. I will continue with the electrons passing through the magnet, returning to the photons later on.

If we take a vertical field and separate out the electrons that move up, and then pass them through a second vertical field, then they move up again. A similar thing happens with down-moving electrons, and with left- (or right-) moving electrons in a horizontal field. So it looks as though the observation of movement (up or down, left or right) is giving us knowledge about the electrons, which is confirmed and shown to be correct by a second measurement. An electron that has moved up is, it seems, like a bar magnet pointing upwards.

The difference between this observation and conventional knowledge of reality emerges in a more complex sequence of magnets (Figure 5). We pass a beam of electrons through three successive magnets, with fields vertical, horizontal, and vertical again. After the first magnet we block off one beam, leaving only the one that has moved *up*. When these reach the second magnet, with the horizontal field, half move to the left and half to the right; of each of these, when they enter the second vertical field half move up and half move *down*, even though originally all of them moved up. The 'sideways-knowledge' provided by the horizontal field has destroyed the 'up-knowledge' provided by the first vertical field. The two pieces of knowledge are not cumulative, but mutually exclusive — complementary.

Could it be the case that the destruction of knowledge is the result of a purely physical interference with the electrons, the horizontal field exerting forces which flip the electrons round, altering in a concrete physical manner their situation? Two arguments show that this is not the case. First, the phenomenon happens however weak the field is. Second, if we fail to take note of which particles move to the left and which to the right, but instead insert further magnets to bend them back again, thus throwing away the 'knowledge' we might have obtained, then when the recombined beam enters the second vertical field *all* the electrons move *up*. In other words it is the mere fact of

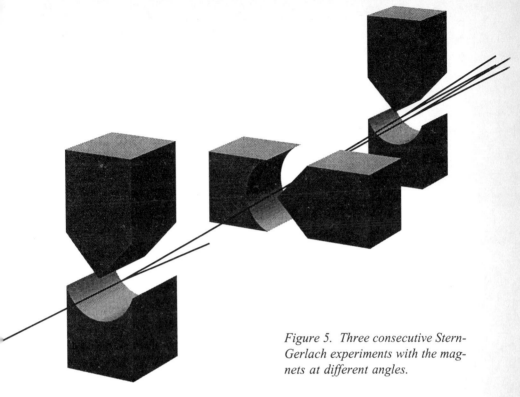

Figure 5. Three consecutive Stern-Gerlach experiments with the magnets at different angles.

getting the left-right information, not the physical interference with the electrons, that destroys the previous information.

This argues that we are not passively getting information about an independent reality, but that we are creating the reality by what we choose to examine.

6. Non-locality

We finally return to the photons and the prism to describe the experiment with the profoundest consequences for our view of the world, the Aspect experiment. Though it uses particles of light (photons) rather than electrons, the arguments are very similar to those of the preceding experiment. We recall that when a photon passes through the prism it either goes straight through or is reflected out of one side of the prism (just as the electrons moved either up or down). On the wave picture of light, a wave is divided into two beams, one passing through, one reflected, with the electrical vibrations in the waves being in different directions in the two beams. Physicists say that the two beams have 'different polarizations,' and the same terminology is carried over to the particle picture as well, with the photons being referred to as having different polarizations according to whether they are reflected or pass through.

Having got these details out of the way, I can now describe the experiment. Many night clubs (and also a local Italian restaurant, a favourite in our family, when it celebrates birthdays among its clientèle) sometimes switch off the main lights and turn on ultraviolet ones, at which all white garments glow blue because of the fluorescent dyes used by detergent manufacturers. The molecules of these dyes absorb light of one colour (in this case, ultraviolet) and then re-emit it a moment later at a different colour wave-length (in this case, blue). Some dyes can be set up so that, when a molecule absorbs a single photon of light at one colour, it later emits two photons in quick succession at another colour. The experiment makes observations on these pairs of twin photons coming from a single molecule, analysing their polarizations using two of the prisms just described, one for each photon, placed on opposite sides of the tube containing the dye, at some distance from it (Figure 6). For the sake of reference, I shall call the two prisms (and the two photons) 'left' and 'right.'

What happens depends on how the prisms are oriented. When both are aligned in the same way, then the same result is obtained from both: either both photons pass through or both are reflected (it cannot

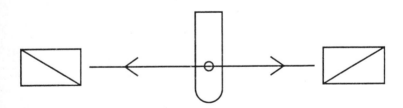

Figure 6. A tube of dye-emitting pairs of photons, analysed by polarizing prisms on each side.

be predicted which of these happens on any particular occasion). In other words, the photons are not acting independently. In itself this is not surprising, since they have a common origin and so might be expected to have similar behaviours.

In view of the picture that we have already built up, however, this is interesting, because it seems at first to contradict the argument that the quantum state is not an absolute reality. Suppose, for example, that one of the photons, the left, say, passes through its prism. Then it is determined that the right must also pass through its prism; there is no choice about what the right one will do, once we know what has happened to the left one. So now there are two possibilities:

♦ The first is that the right photon never had any choice in the first place, in that it was somehow fixed from the start that it would go through. But in that case it must have been fixed from the start that the left would go through, as well, since either both go through or neither goes through. So on this possibility the whole system has a fixed reality which we passively observe, and quantum theory does not apply.

♦ The second possibility is that it remains open what is going to happen to the right photon until we observe the left one. This possibility seems really strange, for it means that *our knowing about the left photon* somehow affects the right photon.

Careful observations of what happens when the prisms are at different angles now shows that the first possibility is wrong: quantum mechanics, with its uncertain reality not fixed in advance, is in fact operating. (The details, while not too complicated, involve a little elementary mathematics: see Penrose, *The Emperor's New Mind.*) So we are left with the second option: not only is the reality of the situation constructed by a dialogue with the observer, but the pair of photons respond to this dialogue in a coordinated fashion, even though they are separated by a large distance. Indeed, the distance is so large that it is possible to choose the angles of the prisms after the photons have left the dye, in which case the photons cannot have time to exchange any information about how they are going to respond. *The two particles form a single whole, dynamically determining its reality,*

even though the component particles appear to be separated in space.
There is an underlying unity of being between the particles which
overrides the separateness of space.

Though this experiment is concerned with the more exotic reaches
of physics, its implications are enormous; what seems a tiny result
about an obscure area of science spreads like a virus through the
whole of scientific thought. The importance of separation in space is
the bedrock of the scientist's view of reality; it is tied up with a view
of space and time as absolute structures on which all else depends.
Yet here is an influence, a kind of harmony, which seems to ignore
spatial distance. Physics is rooted in atomism, the doctrine that
ultimate reality consists of separate particles in the void, that interact
only through external forces and collisions. Yet here are particles that
share an inner unity of being.

Suddenly, we make contact with the ideas on space and time
explored earlier (see Chapter 3). If space does not exist as an absolute
reality, but is just a reflection of the way that things are related, then
the continued relatedness of the particles in the Aspect experiment
starts to make sense. When we say that the particles 'move apart' we
means that some of their relations, the spatial ones, get tenuous; but
other relations, concerned with polarization, remain strong. The nature
of the world is now completely the reverse of the atomistic picture. In
atomism, space is absolute; particles are separated by space and any
relations are external relations between separated objects that have to
be established by communication. In the physics of non-locality, space
is relative; particles — indeed all objects — are internally related and
their separation is only one aspect of their underlying relatedness.

Quantum theory has thus moved us into a completely different area
of thinking. In the first place, it has replaced the reality of atomism by
a different reality of relatedness. But secondly, it has overthrown the
very idea of reality itself, since the nature of things is not something
we observe passively, but something we create in an active dialogue
with the world. Physics itself is now supporting the radical philosophical ideas of Feyerabend with which we started our quest (see
Chapter 1).

Before moving on to the implications of this, however, there is one
more aspect of quantum theory to examine: the use of complex num-

bers. This is the most familiar aspect of the theory to physicists, and yet the least understood at a basic level. It is not too difficult to describe mathematically — I shall do that in the next section — but we do not yet know what it really signifies when we start to look at the wider implications of the theory. While many aspects of quantum theory clearly have big philosophical implications about the nature of reality, this one still seems bound up with the intricacies of theoretical physics, still needing to be unpacked to show its true significance.

7. Complex numbers

I began this chapter by describing waves, particles and the conflict between them, explaining that the resolution of this conflict was to be found through the quantum state, which manifested itself either as a wave or a particle, depending on the questions that were asked in the dialogue with the state, waves and particles being complementary aspects of the state. I now need to say more about how this works, and in particular about the way in which waves appear in the theory, because this takes us further into the difference between quantum theory and the ordinary physics of classical systems.

The key to the idea of quantum states is the fact that they can be combined in the (mathematical) operation of *superposition*. What this means in practice is that, by changing the sort of question one asks, it is possible to regard what was previously thought of as a simple state as having within it several different potentialities. When this happens, the quantum state of the system is described as being in a superposition of the states corresponding to these different potentialities. We saw this in the experiments examined earlier, where a particle entering a rotated magnetic field had the potentialities of being deflected one way or the other, and so was in a superposition of these two states. I explained how this was different from the situation of a dice that had been covered, which might be said to have the potentialities of showing a 6 or a 4, because of the fact of complementarity: the idea that there are different, mutually exclusive, questions that can be asked of the particle, whereas the questions that can be asked of a dice build up to a precise classical reality.

There is another (closely related) way of thinking about the difference between the ordinary probabilities that govern a dice and the mathematics of superposition that operates in quantum mechanics. Superpositions contain information about probabilities, with some extra information as well; and it is this extra information that is responsible for their non-classical behaviour. The probability aspect of superpositions is expressed in the fact that different states can be combined in differing proportions; one can have a state which is 20% of one state and 80% of another, so that when one examines it, the second is more likely to be seen. So superposition is a numerical matter, with numbers expressing different proportions. The 'extra information' is expressed in the fact that the numbers one uses have to be not just simple proportions, like 1/2 or 1/3, but complex numbers: two-dimensional numbers represented not by a point on a line, as are ordinary numbers, but by a point in a plane or by a pair of ordinary numbers.

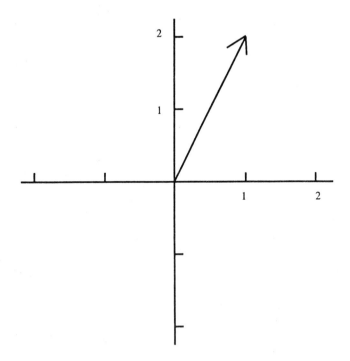

Figure 7. The complex number (1,2) represented by an arrow.

If the symbols ✿ and ○ represent states, then a superposition using ordinary numbers would be represented by a formula like:

$$3✿ + 2○$$

while a superposition using complex numbers would be represented by a formula like:

$$(1,2)✿ + (2,3)○$$

The likelihood of seeing ✿ rather than ○ depends on the magnitudes of the complex numbers involved; the fact that they are complex numbers and not ordinary numbers is only revealed when one starts to make superpositions-of-superpositions!

So, finally, I can explain how waves fit into this picture. *A wave state is a superposition of particle states.* Suppose we represent complex numbers by arrows pointing from the zero complex number (0,0) to the complex number in question. Now suppose we want to describe a wave moving in a fixed direction, and we measure distances from some starting point, in the direction of the wave, marking off points at 1 cm, 2 cm and so on from the starting point. We can denote the state of a particle at position 1 cm by ①, at 2 cm by ②, and so on. Then we can form a superposition of the form

$$→① + ↗② + ↑③ + ↖④ + ...$$

that is, a state where the direction of the complex number moves round as we go from one point to the next. If we now imagine doing this with very closely spaced points, we get the idea of a state, formed by superposing many point-particle states, with a complex number that moves continuously round as one moves along. This is the quantum wave. Unlike a water wave, where the motion of the wave is up and down in ordinary space, the motion of a quantum wave is a motion in the two dimensional space of complex numbers. It is an abstract motion, on the side of the implicate order, not the explicate order. This means that the 'waviness' of the wave is not a fixed part of reality. We cannot think of it as having a definite frequency that we can tune into, unless we explicitly set up an experiment that brings the wave into ordinary space and time.

When the quantum state is thought of as a wave in this way, it is then referred to as the *wave function.* Some books refer to quantum mechanics as wave-mechanics, meaning that it represents the quantum state as a wave.

8. Hidden variable theories

Earlier I examined the objection that the randomness in quantum theory was similar to that exhibited by a dice that was rolled, hidden and then inspected; and this was countered by looking at the strange behaviour of spinning particles (the Stern-Gerlach experiment) which are very different from dice, and the even stranger behaviour of twin photons. These experiments suggest that randomness — lack of pre-dictability — is the key ingredient in all quantum mechanical phenomena, and that this randomness is *irreducible,* unlike that of the dice, whose randomness is due to our ignorance of the detailed me-chanics of its rolling. There are some physicists who would none the less insist that all randomness in physics is actually of the sort shown by the dice: that there are always factors which actually determine the outcome of every experiment, but whose operation is always hidden from us. Factors of this sort are called *hidden variables.*

What the previously discussed experiments have shown is that hidden variables, if there are such things, have to be very strange indeed. They are certainly not simple things like the position and direction-of-spin of particles (as shown by the Stern-Gerlach experi-ment) and they are non-local, acting instantaneously across regions widely separated in space (as shown by the Aspect experiment). In effect, one has to have infinitely many hidden variables, all equally unknown and mysterious, and to have a different one coming into play to explain each new quantum event. This is clearly not science, but a desperate attempt to rescue a mechanical view of the universe, against all the odds, at any cost. Nothing is gained by resorting to theories like these, which are always much more complicated than quantum theory, and which fly in the face of all the evidence to the contrary. The exception to this, however, is the very distinctive use of hidden variables pursued by David Bohm. Unlike those who use hidden variable theories in order, against the odds, to cling to a Newtonian picture, he fully accepted the strangeness of the world of hidden variables that one had to postulate, if one took this course. I shall examine his view in more detail later on (see p.102).

9. The interpretation of quantum theory

Suppose now that we accept that quantum theory is a part of life, or at least a part of the lives of physicists. Even if the rest of us can get along without it, physicists have no real alternative to using the language of quantum theory, in which one talks about a quantum state, mysterious and unvisualizable. The state is interrogated by the observation in a dialogue between the observer and the observed. Or, better (since ultimately there is no distinction between the observer and the observed), in each moment the world is in dialogue with itself. Out of this dialogue, this participation of ourselves in the universe, of the universe in itself, emerges some definite event, lying within the range of possibilities defined by the experiment but otherwise completely unpredictable. What does this mean about the way the universe is? What is going on behind this language? We need to survey some of the very varied descriptions that physicists have used to make sense of this new picture.

The collapse of the wave function

The doctrine of the collapse of the wave function is the view of the physicist who, while prepared to abandon a mechanical universe in which everything is determined, still needs to hang on to a view of the world in which everything that the physicist investigates has a clear reality — the view of the *scientific realist.*

In particular, the quantum state, abstract and mathematical though it is, is regarded as 'real.' On observation there is a real and dramatic change in the state, so that it comes to express the result of the observation. For example, the state of a photon coming from a rotated polarizing prism and approaching another prism aligned vertically is regarded as being in a state which is a superposition of two sub-states: that in which the photon will go straight through the second prism, and that in which it will be reflected. When the photon is actually reflected (say) then its state changes abruptly; the superposition, it is said, 'collapses' into a single state because an observation has been made. When this is coupled with the representation of the quantum state as a wave (see p.95), it is called the collapse of the wave-function.

Von Neumann, the main proponent of this view, described the evolution of a quantum state as being of one of two distinct types: a continuous deterministic evolution in between observations, and a discontinuous, non-deterministic evolution at observations.

The problem with this view is, how does the state 'know' that it is being observed? Some have tried to answer this by putting forward a particular mechanism for the second type of change: Penrose proposes quantum gravity, Gorini has a 'collapse force.' Many writers, noting that what counts for us is the registering of the observation in our brain, introduce ideas to do with consciousness. Penrose identifies his quantum gravity processes with whatever underlies consciousness, while Wigner uses consciousness in a pure form, and Evans uses special properties of complex systems operating through synapses on nerve cells in the human brain. The problem underlying all these attempts is that they fail to grasp the way in which consciousness is a higher level activity, something that emerges in terms of the large-scale organization of the brain, and is invisible at the small-scale level. It is to do with decisions, information, meaning and so we would not expect it to be reducible to microscopic physical processes. The realist approach is evading what the physics is telling us: that the quantum 'explication,' in Bohm's phrase, is a result of a genuine dialogue with the universe, in which truly mental faculties play a part. This is a key area of the whole question of reality, that will be discussed at length later.

The stochastic approach

Another group of authors (for instance, Bellinfante) hold that quantum theory is not applicable to single systems, but only to the statistics of large numbers of systems. Their argument is a bit like those who identify the quantum state with the probabilities of a rolled dice, but it is more subtle.

What physicists always do, they would argue, is to perform a large number of experiments and then compute averages. It is these averages that the quantum state can make predictions about. Thus, they continue, the quantum state is not the state of a *single system* at all; rather, it describes a collection of systems, namely, all the systems used in the many trials that make up a physicist's experiment. Or, more precisely, it describes the ideal case of infinitely many trials, to which the finite number of trials made by the physicist is an approximation. This ideal infinity of trials is called an *ensemble* of experiments. Some statisticians look at ordinary probabilities in this way, holding that when one says 'the probability of this dice showing a 6 is 1/6,' then one is implicitly referring to such an ideal collection of trials, in which a fraction of 1/6 would give a 6. The elegant feature of this approach to quantum mechanics is that, using it, one can show how it is that a quantum state can turn, in the right circumstances, into an ordinary probability distribution. It turns out that by taking an ensemble of measurements, the complex numbers that normally appear in quantum mechanics can be replaced by ordinary numbers, and these can be interpreted as ordinary probabilities. The extra information contained in the complex numbers, over and above probabilities, averages out when one takes a large array of similar instances in an ensemble. This, they would say, is what happens in a measurement: there is nothing mysterious about it, only a well-understood passage from abstract superpositions to ordinary probabilities.

At first sight this seems miraculous. But on closer inspection, it appears to be a 'cop-out.' The proponents of this view declare that they have no way of talking about any individual observation, only about ensembles: ideal infinities of observations. But how can one have a science that, on the one hand, declares atoms and particles to be the most fundamental things in the universe, yet, on the other hand, refuses to talk about the behaviour of an individual particle, only having theories that consider ideal collections of infinitely many observations? If the aim of science is to describe the universe as it is, then we need to be able to describe what happens whenever we engage in that dialogue with the universe that brings to light part of the implicate order. We will be taking up this issue in Chapter 6 in the context of 'broad quantum theory.'

The many-worlds view

This is a radical alternative, mainly developed by quantum cosmo-logists: physicists studying the universe in its early phases when, it is believed, quantum mechanics was applicable to the universe as a whole. The many worlds view therefore takes a cosmic approach right from the start. Rather than talking about a state for an atom, the approach talks about the state of the whole universe, a superstate that encompasses the individual states of all objects in the universe. The seemingly bizarre assumption is then made that, at the level of the universe, the state is real but no collapse ever happens. Instead, when an observation takes place, the state of the universe evolves into a superposition of two (or more) universes, corresponding to the different possible outcomes. This keeps happening continually, so that the state of the universe is actually a superposition of innumerable different 'substates' in each of which a quite different history has taken place. With each observation this 'multiverse' branches into more strands.

This gives a strange way of looking at the observation processes. On the conventional collapse view (see p.97), when I examine an atom the wave-function of the atom collapses into some definite outcome that I measure. On the many-worlds view, when I examine an atom the universe as a whole branches into two (or more): each branch contains a different copy of myself, and each copy is observing a different outcome for the experiment. This view still does not answer the question: how do we know when there is an observation taking place? Observationally it is indistinguishable from the collapse view, but adds a dubious metaphysics to it.

Lockwood's 'Perspectives'

This is a variant on the above. Like some of the 'collapse' approaches that I have described, it makes consciousness the key to the emerg-ence of a definite outcome to an experiment, and so it introduces a theory of consciousness into quantum theory. According to Lockwood (in *Mind, Brain and Quantum: the Compound I),* consciousness is a perspective, a view from a particular vantage-point, so to speak, on the multiple superposed universe. The multiverse is in itself a homo-geneous whole, without any clearly separated branches. Within any evolved universe containing superpositions of observers there are many different consciousnesses, each of which presents a view of the universe in which only one definite situation obtains. It is the per-spectives of consciousness that define different strands within the multiverse. Thus we do not have to answer the question, what consti-tutes a measurement: it is not the measurements that are responsible for separating into strands. Instead we must ask, 'what constitutes a conscious perspective?'

Lockwood's idea of a perspective comes from his analysis of consciousness which I shall describe in more detail later on (see p.139). The problem here is that the theory of consciousness that is re-quired is only very briefly indicated by Lockwood. All the difficulties are shifted off onto this unknown theory of consciousness that still needs to be developed. I shall, however, return to this later when I have examined consciousness in more detail.

Veiled reality

Bernard D'Espagnat, in his book *Quantum Theory and Reality* (1983), moves towards a position that seems in many ways the most honest expression of what emerges from the experiences of quantum physi-cists. To begin with, he reviews the traditional approach of scientists to reality by quoting from a textbook by Messiah:

> At the start of every scientific enterprise an assumption is
> made that nature possesses an objective reality, independent
> of our sensory perception and of our means of investigation:
> the object of a theory is to give an account of this objective
> reality.

D'Espagnat calls this the 'Postulate of Physical Realism.' He con-cludes, after analysing experiments such as those we have described above, that one cannot accept this postulate in its full form. On the one hand, the whole scientific enterprise depends on the idea that the physical world really is out there: it is not a complete construction of our imagination, but we are hitting up against something that responds to us as an 'Other.' Science depends on the possibility of being wrong, which implies something with some measure of independence from ourselves. On the other hand, this 'something' is not open to our inspection, as was the case with classical physics. It is a reality, but it is veiled from our gaze. D'Espagnat says that 'If I want to retain my realistic requirements, I am thereby compelled to embrace a non-physical realism, which might be called a *theory of veiled reality*. ... I understand [this] to mean any realism that does not satisfy the hope ... which the Postulate of Physical Realism summarizes.' In Bohm's terminology, examined below, the implicate order is real, but it remains implicate and must not be confused with the explicate order revealed to our senses.

It might be said that what D'Espagnat is advocating here is an attitude, more than a theory. If something is always veiled, never the object of our direct awareness and never accessible to indirect observations with instruments, then in calling this 'reality' we are making a declaration of faith, an assertion that, appearances notwithstanding, it is something that counts in the world and we ought to build our understanding of the universe upon it. It is an attitude that does make a difference to what the scientist does. Believing in 'veiled reality' the scientist will direct the construction of theories towards it and will be happy to introduce theoretical speculations about what it might be like. We need to remember, however, that this attitude leads us into quite strange uses of the word 'reality,' when what is forever unobservable may be called reality, and what is the most immediate aspect of awareness may be called subjective and unreal. On such a view, what is most obscure is most real. Have we not somehow turned the meanings of words on their heads? This suggests that we should not rest content, either with veiled reality, or with the closely related implicate reality of Bohm, which we examine next.

The implicate order

The late Professor David Bohm, with his collaborator Basil Hiley, has developed a theory — strictly speaking an extension rather than an interpretation of quantum theory — that is more detailed and complete, both in its mathematics and its philosophy, than most of those I have described so far. It has its origins in an approach similar to the hidden variable theories described earlier (see p.96), in that it is supposed that there is more in the world than just the wave-functions, but this 'more' is not observable directly. They assumed that the physical world consists of both waves and particles in interaction. Thus in the Stern-Gerlach experiment there is both a quantum mechanical wave passing through the magnet and spreading into two separate beams going either up or down, and also particles that have a definite position in space. The particles do not, however, obey the laws of Newtonian mechanics, but are instead 'guided' by the wave. Depending on how they enter the apparatus, they may find themselves in the part of the wave leading up, or the part leading down. If a great many particles are sent through the apparatus, then they accumulate where the wave is strongest, giving the observed results. There is much to commend this theory. The individual particles, entering the system in a different position at random each time the experiment is performed, explain both the overall randomness of the results and also how it is that a definite result is obtained on each individual occasion. The waves, on the other hand, explain the more radical features of quantum theory — its non-locality, and the participatory role of the observer who defines the structure of the waves by the nature of the experiment that is set up. There is no need to assume that the wave-function collapses, because if it were to separate into two distinct parts, which never subsequently recombine, then the particle must always be found in one part or the other, where it will stay. The part of the wave where the particle is will thereafter be active, while the other part will be inactive and can henceforth be ignored, without having to assume that it disappears in a collapse.

Bohm and Hiley, in *The Undivided Universe* (1993), stress that theirs is an *ontological* theory, making assertions about what actually *exists,* whereas in the conventional theory the state merely expresses a potentiality for what might exist. It would seem from this that Bohm's implicate order of waves and particles is regarded as reality, even though it is not a fixed reality that we passively measure as external observers, but a dynamic whole in which we are active parti-cipants. Yet the question has to be asked, why is one justified in ascribing ultimate reality to Bohm's complicated system of particle dynamics, when it must by its very nature remain for ever unobservable, and when its detail — even its entire structure — is subject to replacement by a new system whenever the prevailing scientific theory changes. Bohm and Hiley themselves describe such a replacement, in which a deeper analysis shows that the particle component must be replaced by a wave component for full agreement with physics.

Declaring the implicate order to be some kind of ultimate reality suffers from the same problem as does D'Espagnat's veiled reality: it starts to turn the words on their heads, ascribing reality to that which is most remote from our experience. Unless — and this will be a key proviso — there turns out to be a realm of experience which is both experienced as supremely real and also underlies the physical world. This is the possibility that we will be examining next.

10. Quantum reality

We have seen that quantum theory has dramatically altered the Newtonian view of reality (the atomistic picture of particles and the void) in several different ways, all of them grounded in very specific physical experiments.

First, it has undermined the basis of this picture by showing that the universe cannot be thought of as made up of separate particles, but, because of non-locality, is interlinked into a single whole. Particles are no longer isolated entities, related purely externally, but become extended waves, manifesting themselves everywhere and internally linked with each other in a single web.

Second, it has altered the very foundations of any sort of concept of reality, by suggesting a way of thinking of the universe in which reality is not something that is passively 'out there' waiting to be observed, but which emerges from our own dynamic participation in the universe, and from the choices that we make in so participating.

Third, it has rendered the whole notion of 'reality' fluid, by showing that in this complex tissue of events that are continually coming into being, as a result of continual creation and participation, there are no clear divisions between theory and reality, or between reality and appearance. Better, it has underlined the fact, already demonstrated by Feyerabend, that there never were such clear divisions, and has blown away the Newtonian pretence that such divisions were possible.

We have touched on Bohm's presentation of a world view that encompasses participation and non-locality, while still maintaining that there is a specific reality behind it all, in the implicate order. Powerful though the system is, however, it depends on an ultimate rejection of Feyerabend's critique, and a rejection of any similarity between Feyerabend's historical engendering of scientific theories and the quantum mechanical participation of the observer in the creation of reality. In this book I am arguing that postmodernism and quantum mechanics are both delivering to us the same message: reality is not something fixed, either in its general nature or in the specifics of what takes place; but neither is it determined purely by human whim. Every situation is both indefinitely open to the potentiality of the creative act, and yet also indefinitely constrained by the historical weight of all that has gone before.

Bohm's view may rescue a scientific concept of reality from the critique of quantum theory, but it cannot in itself combat the deeper attack from postmodernism. What it does is to point to the need for a conception that is powerful enough to withstand the combined attacks of postmodernism and quantum theory. We will be moving towards such a theory in the following chapters, where I shall justify an approach in which reality does not lie in an absolute physical realm, as Bohm would have it, but in the relationship between the active subject (in our case, the human subject) and the whole context in which that subject finds itself.

Chapter 5
Consciousness

The question of consciousness has cropped up repeatedly in the previous chapters. First, it is present in the atomistic approach of the seventeenth century, when Descartes divided the world into matter and spirit, making matter the domain purely of geometrical properties, while most of the qualities that we are aware of in the world arise from the spirit, the seat of consciousness.

Secondly, it arose in quantum theory, where several writers have, in different ways, invoked human consciousness as an essential part of the physical system of the world. For these, human consciousness either plays a part in the collapse of the wave function (see p.97) or in separating out a multiverse into different perspectives (see p.100).

Since consciousness is so closely linked, both with the traditional scientific view of reality and with the quantum view that might replace it, we cannot evaluate these views of reality without studying the place of consciousness in the picture. But there is an even more fundamental reason for looking at consciousness. I explained at the outset that the primary sense of 'reality' is that of 'the way the world is,' (see p.20) and that in common every-day usage, before the scientist gets to work analysing the idea, it means simply what we are aware of when we are not deluded or imagining things. This 'simply what we are aware of' is also the primary meaning of consciousness. The starting point of every inquiry must be our own awareness. If we ask about the 'me' element in this awareness, then we start exploring the realm of personal consciousness; if we ask about the totality of our awareness, and start comparing it with the awareness of others, then we move towards distinguishing illusion and so defining reality. Consciousness — the fundamental givenness of the world to awareness — is the starting point for any discussion of reality. So what is it, and how does it fit into the different scientific approaches?

1. Consciousness in the Newtonian world

The Newtonian world view, following from the atomism of the Greeks and Lucretius, and the philosophical revolution initiated by Descartes, had created a division between physical reality — the world of matter characterized by geometrical properties, devoid of colour and other human qualities (so-called secondary qualities), explicable in terms of atoms — and the secondary qualities that were added in human consciousness, which accounted for the bulk of our experiences. Descartes would have said that some of these qualities were added by the mechanical workings of the brain, others by the soul. Today scientists would say that all these secondary qualities are introduced by the brain, rather than the soul, because their aim is to explain everything without reference to unknown spiritual influences such as the soul.

What these scientists, from Descartes on, are doing is twofold. First, they tried to explain certain observations obtained through carefully planned experiments; these gave rise to scientific theories and these in turn to the whole wealth of technology that is linked inseparably with science. Second, they tried, through their building up of a philosophy of nature, to explain in general terms the totality of our awareness; to explain the world as it floods in upon us every second of our lives. These two sorts of explanation were at first intertwined. Both were described as 'natural philosophy,' which later became separated into 'science' and 'philosophy.' And the French *expérience* means both 'experience' and 'experiment.'

Now, when the scientist lays claim to declare the nature of reality, it is this second activity that is being referred to. The scientist is putting forward a picture that is claimed to underlie the whole of our awareness; and where it seems to differ from what we are immediately aware of — when it speaks of atoms or a moving earth, though we can see no atoms around us and cannot see the earth move — then this has to be explained in terms of the inadequacies of our senses, or in terms of illusion. For Descartes, all the secondary qualities (colour, smell and so forth) are illusions and the only true physical reality is geometry.

If the claim by science to pronounce on reality is to be justified, then this second scientific activity of explaining the totality of awareness, of consciousness, has to be carried through completely; or at least sufficiently completely to persuade us to accept it. Unless this is done, science may be able to predict the result of contrived experiments, or to produce technological marvels, but it can offer nothing to enable us to understand our place in the universe. So, for a science based on Descartes' view to be successful, it needs to explain how it is that the external world of atoms, together with the processing of our brains, gives rise to everything that we are conscious of. It is not enough just to construct a self-consistent theory of the results of physics experiments, however great an intellectual triumph that may be, however useful it may be in order to control certain parts of nature and to develop technology, if the world that we live in from one day to the next has no explicable relation to the scientific world of experiments.

A full scientific explanation of our awareness must include not only an account of the world of atoms and geometry, but also an account of how this world becomes translated into our awareness of colour, beauty and so on. It is not enough just to describe the atoms and geometry and then baldly to state that we form the illusions of colour and smell; there has to be a link between the external geometrical world and the internal conscious world, or else the project fails. This was precisely where Descartes himself failed. He had a quite separate physical world and an immaterial soul, and postulated that the pineal gland linked the two. But this was in no way an explanation; it offered nothing to explain why certain physical actions should give rise to the particular sorts of awareness in the soul that they did. If science is to speak of reality, the circle must be closed through an explanation of sensation and consciousness.

Consciousness and action

So far we have concentrated on that part of consciousness that is our awareness of the external world, of the green trees and the blue sky; in other words, on perception. We shall return to perception, and the problems that it poses for a mechanical account of nature, later. But there is another part of consciousness, of our awareness, that has a different form: our planning, judging, deciding, acting. This was of particular interest to Descartes, because to him it seemed that these activities were particularly human, and for him humanity was the main topic of interest. There is no reason to suppose that there is a single 'thing' called consciousness that is responsible for both perception and action, but that assumption is so frequently made that it is rarely questioned.

Thus Descartes set out to try to explain the basis of consciousness by tackling the origin of conscious action. And the key point of his account was the idea that only part of human action could be explained in mechanical terms; for the rest one needed a soul, interacting with the body through the pineal gland. Though the idea of the pineal gland was a non-starter, there is much about Descartes' thinking that is relevant today. In particular, it is relevant to modern attempts to give a purely mechanical account of consciousness in terms of the operations of the brain. Can there be a purely physical account of reality, including consciousness (and, in particular, conscious action), or must one bring in some extra, non-physical entity, whether it is called a soul, mind, or whatever?

Is mechanics sufficient?

We have seen how Descartes attributed a great many mental functions (perception, memory, emotions and the details of moving the limbs and organizing the body) to the purely mechanical working of the brain. He imagined the brain to operate by a system of very delicate hydraulics, which eventually became amplified sufficiently to move the limbs. He was, however, very clear that there were certain human functions that could not possibly be explained in this way:

> We certainly can conceive of a machine so constructed that
> it utters words ... corresponding to ... a change in its organs
> (e.g. if you touch it in one spot it asks what you want of it,
> and if you touch it in another spot it cries out that you are
> hurting it). But it is not conceivable that such a machine
> should produce arrangements of words so as to give an

appropriately meaningful answer to whatever is said in its
presence, as even the dullest men can do.

For this, Descartes argued, the soul is required; and so the whole
human being is explained through a combination of soul and body.
Now the soul, according to Descartes, is simple, in the sense that it
cannot be analysed into parts or components: it just is. So any further
analysis of the internal workings of the soul is impossible, and scientific enquiry stops at this point. The soul of Descartes is not a
scientific solution, but a prohibition on any further scientific investigation.

This then raises two distinct questions. First, as we consider more
and more sophisticated sorts of human behaviour, is there some point
at which a mechanical explanation has to stop, at which something
totally different has to be considered in order to make sense of the
human being? Second, is it the case that, here or at some other point,
not only does the mechanical explanation have to stop, but all scientific explanation, of any kind, has to stop? The answers will depend, of
course, on what one means by 'mechanical' and what one means by
'scientific'; but I for one would be deeply suspicious of any absolute
'yes' to the second question, of any view that held that there were
absolute barriers that for all time blocked any further scientific
exploration. Certainly there are times when exploration has to be
carried out with the deepest reverence for the values and qualities of
what one is exploring; there are times when the only and the right
response is to fall silent before a mystery that can be entered but not
understood. But to say that there are 'no entry' signs that forever
block the human journey at certain points is something quite at vari-
ance with the nature of consciousness as we know it, which is to
explore without limit.

Whether or not such exploration can, or should, be called scientific,
and what the nature of science actually is, will occupy us later. For
the time being, let us revert to the first of the questions, as to whether
there is some point at which mechanical explanations cease, a point
where we must bring in something new, which we might choose to
call a soul. From a modern point of view, Descartes' claim that a soul
was required in order to perform many operations that a machine

could not possibly do seems rather suspect. With the coming of computers we now have a much wider range of examples before us than the few clockwork automata that strutted in glass cabinets in Descartes time, and we can imagine that machines could do a great deal more than he thought. Information technologists are working on such things as airline booking systems which could respond appropriately to spoken requests for flights, ask appropriate questions, and take (hopefully) correct actions. Though few of us would be willing to trust our holidays to such a system tomorrow, we could imagine a time when this might be working smoothly.

Although this does not amount to giving an appropriate answer literally to 'whatever is said in its presence' (since the system's vocabulary is limited to airline flights), there are many humans whose vocabulary is rather limited, and we do not as a result deny them consciousness. Is there anything, we must ask, which human beings can do which *even in principle* no machine could manage? If so, then we would have established the need for something non-mechanical, and reality would be more than the machine that modern scientists often claim it is.

If one could find such a task, which no machine could accomplish, then it would provide a test that would distinguish between a machine and a non-machine. One could imagine communicating via a display unit with some unknown source of messages in an inaccessible room, faced with the problem of determining whether the source of the messages was a person or a cunningly programmed computer. What would be needed would be a question, a request to perform some task, that could only properly be answered by a human being and not a machine. This would prove that Mechanics was not adequate to explain conscious action.

The Turing test and Gödel

A test of this form is called a Turing test, after the mathematician and founding computer scientist Alan Turing. A long debate has been carried out, involving computer scientists and brain-scientists, as to whether such a test can be devised. One of the most recent contributions has been from Roger Penrose, who has claimed that a Turing test can be devised in the realm of mathematics, by using what are called 'undecidable propositions.'

To explain this, we need to look at the nature of modern mathematics. The subject is based on the notion of *proof:* a proof in any branch of mathematics is a formal argument starting with axioms that have been agreed as the appropriate basis for that part of mathematics, and where each step in the argument follows rigorously from previous steps by the application of precise logical rules.

Learning mathematics at a university is partly a matter of learning how to construct mathematical proofs. Needless to say, there is a big difference between knowing something and proving something: 'knowing' applies to the world of the senses, proving applies to abstract mathematical systems; and there may be little connection between the two.

Mathematical logic is a branch of mathematics concerned with, among other things, studying the nature of this idea of proof. It usually focuses on the logic of proving simple statements in arithmetic, such as 'if you add together two odd numbers, the result is an even number.' You can probably convince yourself that this is true by trying it a few times; in fact this is a statement that can be proved to be true, and so one can be sure that it will always be the case. Much more interesting are statements like 'there are an infinite number of pairs of primes differing by two.'* This seems to be true, but nobody has found a way to prove it. Until 1931, mathematicians tended to assume that if a statement like this was true, then there must exist a proof of it, if only one is clever enough to find it. But in this year the world of mathematicians was thrown into turmoil when Kurt Gödel showed that in any sufficiently complex branch of mathematics (it just has to be sufficiently rich to include elementary arithmetic) there were always statements that could neither be proved false nor proved true. The statements that Gödel discovered were worse than the ones just quoted. They were not merely ones for which nobody had yet found a proof; they were statements for which Gödel could show that there was no proof, or disproof, to be found.

---

* A prime is a number like 2, 3, 5, 7, 11, 13,... that cannot be divided by any other whole number except 1. Examples of pairs of primes differing by two are (3,5), (5,7), (11,13), (17,19), ...

Statements that could neither be proved false nor proved true are called *undecidable propositions*. Their appearance was a shock because hitherto mathematicians had assumed that if a proposition was true then it must be possible to prove it to be true, and similarly if it is false. In retrospect this assumption was completely unreasonable: it smacks of the idea that human beings are omnipotent when it comes to logical processes. Maybe the idea that all mathematical statements could be proved arose from the argument that if mathematics was a human invention, then human beings ought to be able to understand that invention fully — an argument that begs a lot of questions.

The interest of undecidable propositions, from the point of view of consciousness, is that if one looks at the argument used by Gödel to show that a proposition was undecidable, this argument also establishes the fact that this particular proposition is true! The mathematician can, so to speak, step outside the formal system of logic that characterizes whatever branch of mathematics he or she is dealing with, and from this higher viewpoint 'see' that a proposition is true, even though there is no proof of the proposition within the mathematics itself. So the mathematician, as a human person, seems to be doing something that goes beyond the capabilities of formal logical reasoning. Might this be a clue to some sort of insight: that a human can perform, but not a computer?

At first this seems to be a rather tenuous sort of argument, being restricted to mathematics — something largely of interest only to mathematicians. It is, however, much more general than this because any computer can be thought of as carrying out some sort of mathematics (even if the result of the mathematics is presented to the user as a picture on the screen, a poem on a piece of paper, or a piece of pseudo-Bach played through a loudspeaker). So there will be some logical operations, or problem-solving tasks, which, according to Gödel's arguments, the computer cannot do, because they correspond to undecidable propositions in the computer's mathematics. The mathematician, however, has the capacity of getting outside the mathematics and so solving the problem that the computer cannot solve. Penrose, in *The Emperor's New Mind,* takes this as evidence that the brain can do things that a computer cannot do. Determining the truth or falsity of undecidable propositions gives us a Turing test for distinguishing human beings from computers. Admittedly, it is a very esoteric and technical sort of test, but the point is that the test exists, if Penrose is right.

Unfortunately this argument (like that of similar arguments produced earlier by Lucas and others) is based on a fallacy that keeps recurring in many arguments about consciousness. We might call it the fallacy of the shifting vantage point. We have to be very careful to distinguish the operations of 'me thinking about me' and 'you thinking about me': the objects are the same, but the vantage points are different. When we were talking about problems the computer could or could not solve we were speaking from the vantage point of the computer, able only to look at itself. When we brought in the mathematician, we shifted our vantage point to someone outside the computer, able to look in and examine its parts from the outside; from that vantage point, new things become possible. But the new ability to solve problems is not due to the fact that the mysterious insight of the human organism has been brought in, but it is just due to the change in vantage point. We could equally well have brought in a second computer, programmed so as to be able to analyse other computers and determine their undecidable propositions. This is a completely logical and programmable process, except that it is a process that can only be carried out on a second computer, not on the computer that is itself being analysed. Gödel's argument does not say that the process of deciding undecidable propositions is non-logical (it is one of the great triumphs of pure logic); what it says is that it is a process that requires a larger system than the system within which the proposition is first formulated.

Let us try to pin down the issue more precisely. What is the difference between the operation of the computer, which we think of as 'mere' machine, and the operation of a human being that we think might involve powers of insight that a machine cannot reach? The point is that the computer (or, to be precise, all the computers that one can easily buy off the shell) work by obeying sets of completely fixed rules. Though it may often seem that they are capable of sheer bloody-mindedness, in fact they are always responding according to strict rules that somebody has fed into them. This is their strength (reliability) and their weakness: they cannot have strokes of genius when they decide to tear up their programmes and do something totally new. The set of rules (the programme, in a sense) that governs a particular set of actions by a computer is called an *algorithm*. What we are trying to do, therefore, is to distinguish algorithmic (rule based) operations from non-algorithmic operations.

Penrose is arguing that, in proving Gödel's theorem, the mathematician moves out of the system that is being investigated into a larger system; and in this larger system it can be shown both that there is an undecidable result (for the first system) and that this result is in fact true. Thus the mathematician seems to achieve something that goes beyond any given system. To see the fallacy of the argument, however, one has only to read the proof of Gödel's theorem to see that it is itself based on an algorithmic system. One could programme a computer quite easily to accept as input the rules of some logical system, and to churn out as output that particular undecidable true result that Gödel's theorem gives for this system of logic. Such a computer would be using a high-level algorithm to go beyond the low-level algorithms of the logical systems that one fed into it. It could be the same with a human mathematician, who could be using high level algorithms wired into the computer of her brain to go beyond the low-level algorithms of the theories she was studying. (Of course, if this were the case, then some even higher intelligence could apply Gödel's theorem to the mathematician's brain, and deduce that there was some result that she could never prove, even though it was true; but she could have no conception of what this result could be.)

Because of this fallacy in Penrose's argument, and in similar arguments, so far these purely logical attempts to find a Turing test have failed. So let us turn to more commonsense ways of trying to find the Turing test that would reveal the essentially non-mechanical nature of a human being. Imagine that we are presented with a computer screen on which messages appear for us to reply to, and that we are trying to determine whether the screen is connected to a computer, or to a person with another terminal in the next room. Few of us, when faced with the problem of deciding whether the messages flashing onto a screen were produced by a computer or by a human, would plunge into pure mathematics. We would probably start probing purely human experiences. Could it tell us what it was like to be in love? To meditate alone through a dark night? To smell the first flowers of spring, or to eat melon and strawberry? If it gave satisfactory answers to these questions, then the evidence for a human being would be high. Whereas, if it showed a tendency to discuss what it was like to calculate *pi* to ten million decimal places, or to solve partial differential equations in ten dimensions, then evidence would accumulate for its being a computer.

On this view, a human being is characterized by a range of experience that is fixed not by logical restrictions, but by the particular sorts of physical organisms that we are. It is the body that determines the uniqueness of our minds. We have a paradoxical reverse of Descartes' position. He was arguing that the most completely mental functions, the higher forms of reasoning, were necessarily impossible to carry out with a machine. We have reached the suggestion that, if by 'machine' one means a computer constructed out of silicon and plastic, then it is the parts of our mind most connected with the body that show us to be different from this sort of machine. We will return to Descartes later; but let us spend a moment longer with Turing.

Does the use of questions linked to bodily senses give us a Turing test? Unfortunately, the answer has to be, no. In principle it would be possible to imagine a computer so large and complex that it could be programmed with all the data necessary to simulate all the chemical processes in a human being, with a programme that was restricted so that only this was possible. A programme that functions in this way is sometimes called, by computer scientists, a 'virtual machine': the real machine is the large computer that is running the programme, the virtual machine is the programme that is pretending to be a human being. So if we allow vast computers simulating human beings as virtual machines — and there is nothing in the rules of the Turing test to forbid this — then this bodily attempt at finding a Turing test will also fail. Probably we will have to admit that there simply is no Turing test.

We must now reflect back on how this affects the original argument of Descartes. He was looking at what a human being could do, the sounds it could utter which we take for speech, and arguing that some of these performances could not be done by a machine, where by this he had in mind a device operated by clockwork or hydraulics. We have now shed doubt on this from several directions. First, Descartes' higher mental functions (and he would certainly have included mathematics) do not seem to be ways of getting behaviour that is intrinsically non-mechanical. Second, it is not even necessary for an imitation human being to be constructed of the same sort of materials as we are, for it to produce the same sort of behaviour, since a sufficiently large computer could simulate a human being as a virtual machine. Thus the particular argument used by Descartes seems to fail in the light of our greater knowledge of the capabilities of machines; the argument gives no grounds for concluding that the human being has to be thought of as having a soul that is separate from the body.

This would be the position of most scientists today. According to this, a mechanical explanation of the human being is completely possible, and there is no boundary at which either a mechanical explanation, or, more generally, a scientific explanation, has to stop. On this view, therefore, there is no reason why the whole of the appearance of the external world, the whole of reality as we actually experience it through our senses, cannot be explained through mechanical processes in the human brain, with no recourse to a soul.

2. The problem of qualia

So far we have examined the argument against mechanics from the point of view of behaviour; of the way a human being functions when viewed objectively, from the outside. This was the argument that Descartes used, and the argument surrounding the Turing test and Gödel's theorem. But a quite different set of problems arises when we look at consciousness from the inside, from the vantage point of our own awareness. It is these that pose the real threat to a purely mechanical account of consciousness.

At this point, as we enter a new area, not considered explicitly by Descartes, we are again bedevilled by the fallacy of the shifting vantage point. The difference of vantage point, between looking at human behaviour from the outside and examining our own awareness from the inside, is crucial, and raises some of the most disputed philosophical questions. Each person is trying to make sense of her/his world, a world, as I stressed many times, that consists of objects, colours, smells, emotions, decisions and the rest. When I discuss my world with someone else, trying to compare it with their world, come to the conclusion that there are some things — precisely those things that I have learned to call external, in the realm of not-me — on which we more or less agree; and other things — the internal things, in the realm of me — which seem private to each of us.

Another common way of expressing the difference is by calling the things we agree on 'objective' and the things that are private 'subjective.' Now, we have already encountered a similar distinction in Descartes' treatment of reality, between primary qualities (position, shape and so on) and secondary qualities (colour, smell and so on). How do all these different classifications — internal/external, subjective/objective, primary/secondary — match up? Is it the case, for example, that secondary qualities are internal and subjective? To answer this, it is best to take a particular case: suppose I and a friend are looking at a red curtain and that we compare notes, talking to each other about what we are experiencing. There are some accidental possibilities for disagreement: some languages classify colours quite differently from English, so that if my friend comes from such a linguistic background there may be confusion about translating the word 'red'; or the lighting in the room may be coloured, reflecting different appearances to each of us. But assuming the lighting is good and that we both come from similar linguistic backgrounds, we will agree with each other, when we talk about it, that what is in front of us is a red curtain; which suggests that we are dealing with an objective description. We might, of course, be sharing a common hallucination, but the likelihood of that is reduced as more observers are brought in to examine the matter from different viewpoints. But what exactly is it that is being declared objective? Is it the totality of that experience which I am referring to in the brief words 'red curtain'; or is it just the verbal description that is objective? When we agree verbally that we are seeing a red curtain, I am likely to jump to the conclusion that my friend's experience is identical with my experience, because we choose the same words to describe it (we all tend to start from the assumption that everyone else is like ourselves); but such a conclusion could well be shaky. It could be, for example, that a part of this total experience is really shared in common, while another part is something that is idiosyncratically my own. How does the subjective/objective classification fit in with this?

This brings us to the vexed question of what are called *qualia* (plural; the singular is *quale)*. A quale is an aspect or component of an experience that cannot be broken down into any smaller or more elementary components or described in any terms other than itself. The standard example is a colour, such as the redness of our curtain. Whereas the curtain itself can be defined in more basic terms (fabric, threads, used for hanging ...), it is not so with redness. Although I can define some colours in terms of others (purple, for instance, as between blue and red) this process is not really decomposing into more elementary terms, since there is no unique choice of, for example, primary colours, on which to base all other colours; it is just an expression of the relationships that exist between colours, all of which are elementary experiences in their own right, as far as our raw experience goes (we are not talking here about the physiology behind those experiences, which can indeed be decomposed into a particular set of primary colours). And although we can talk round our colour descriptions in order to check that the two of us are using the words consistently — saying, for example, that red is the colour of British post boxes and fire engines — this is not *defining* the experience of red. When I say 'red' I am not thinking of post-boxes or fire engines, but of the pure experience of redness. Redness is not made up of the more elementary components of fire engines and post boxes. Rather, it is the other way round.

To avoid misunderstanding, I should make it clear that I am not saying that all experiences can be reduced to qualia, or that qualia are somehow more 'fundamental' than other experiences. It is just that one of the most striking things about our world is the presence of these qualia. A world without them would not be the world that we live in, and so there is a need to understand what they are.

Return now to the question of the objectivity of my experience of the red curtain; and consider whether or not the quale of redness is an objective property. In conversation with my friend, I might probe more closely to check that she was using the word 'red' in the same way as I. We could agree that it is a property shared by post boxes and fire engines; but this is not, as we have seen, a definition of the quale red. It establishes that there is a quale that she experiences in the presence of post boxes and fire engines, and that there is a quale that I experience in the same circumstances, and that both of us have been taught to call our quale 'red'; but are the experiences actually the same? On this hinges much of the debate about the nature of reality.

Many philosophers would at this point call 'foul.' Daniel Dennett, in *Consciousness Explained* (1991) takes a view that rules out any discussion that could possibly get us to this point. In effect, he forbids me to consider my own experiences, saying that it is the business of philosophy and science to consider only other people's experiences as they are revealed through verbal reports. He calls this standpoint heterophenomenology. In the conflict that can arise between different viewpoints, he consistently chooses the viewpoint of the external observer restricting attention to the words spoken and their explicit meaning. The viewpoint is not to be confused with behaviourism, where words are seen just as a particular sort of mechanical noise, to be explained in mechanical terms along with all other sorts of human behaviour. For Dennett the words that people utter are taken completely seriously, but from a consistently external perspective. From such a perspective all reference to my total experience is debarred from the outset, and with it the notion of a quale. 'Red' for Dennett is something to be determined from linguistic analysis of people's use of the term, and that is an end to it. But in doing this, Dennett forbids me from carrying out precisely what I have identified as being the goal of human enquiry, namely exploring my understanding of my world. It makes life a lot tidier, and avoids a lot of philosophical problems, but it is just as arbitrary a step as that taken by Descartes when he ruled that the soul was simple and incapable of being analysed in any way. Both steps draw an arbitrary boundary to the human journey of understanding.

There is another, more powerful, objection to this talk of qualia. I asked earlier, whether my experience of red was the same as my friend's experience of red. Many philosophers would argue that a statement like this only has meaning if there is some way in which I can either verify that it is the case, or that it is not. To understand the meaning of something, they would say, is to understand how one would set about verifying it, and if no verification is possible then there is no meaning. In the present case, there would appear to be no way whatever in which I could determine the question; I cannot 'get inside my friend's head.'

There is another way of posing this objection which makes it even stronger. I have been considering the proposition that two qualia, one of mine and one of my friend, are the same. But there is something rather strange about the word 'same.' Does it have the meaning that it has in: 'is this the same apple that you picked yesterday'? Presumably not, because an apple is a material object that can be tracked to check that nobody removes it and substitutes a similar one, quite unlike qualia. In the case of qualia, the relevant example would be: 'the colour of the curtain is the same as the colour of that apple'; but in this case I am comparing two of my own qualia, not one of my own with one belonging to someone else. While I have a perfectly good idea of what 'the same' means for my own qualia, why should the phrase 'the same' have any meaning at all when we compare different people?

An example from geometry might make clearer this idea that sameness can break down and become meaningless when we move to different vantage points. Suppose I am pointing out an interesting bird to you; you are standing next to me, and I urge you to look in the same direction as I am pointing in. There is no doubt as to the meaning. But now suppose you are in Beijing and I am in London, and we are connected by a television link. If I were to urge you to look in the direction that I was pointing in, you would be perplexed: did I mean that you should try to relate directions on the surface of the earth in Beijing to those in London (and if so, are you to do it by somehow projecting one side of the globe onto the other, or by using compass directions), or should you look in the direction in which I appear to be pointing on your TV monitor (which raises the question, do I know correctly the orientation of your TV monitor). You would probably merely conclude that I had got carried away and failed to realize that words that make good sense when we are standing next to each other fail to make sense when we have very different vantage points. It could be the same with qualia. If they belong to different people then there may simply be no meaning to the question of whether or not they are the same. But that does not imply that the concept of a quale has no sense. There is no clear meaning to directions in Beijing and London being the same, but the concept of 'directions' makes perfectly good sense in each one separately.

The objection that the phrase 'the same' may be meaningless is a powerful one, and I shall only be able to counter it later on, when we will have built up a picture of the world, of perception and reality, based on the insights of quantum mechanics.

I propose accepting the need to seek an account of the fullness of each person's experience, an account that must reject the prohibitions of Dennett's heterophenomenology and squarely face the problems posed by qualia. These qualia, such as the redness of the curtain, are labelled by us as external, as being in the realm of the not-me. But they are not fully objective: they cannot be captured within the verbal definitions that are available from the viewpoint of an outside observer. None the less they are a fundamental part of my world, and as such they call for an account. I want to look in turn at the two extreme possibilities: one, that qualia are, as Descartes might have held, purely added in the brain; the other, that they are completely real intrinsic properties of the objects that we perceive. Neither will be acceptable, and we will have to move on to a more complex view of reality in which the qualia reside in the relation between the me and the not-me, and not in either separately.

3. Mechanics versus mind

To separate out the strands of argument, I shall follow the example of Galileo in presenting an imaginary dialogue between Frodo the physicalist, who thinks that everything can be explained by nerve impulses in the brain, and Minnie the mentalist, who thinks that there is more to it than that.

FRODO. There are people like Dennett who hold that qualia do not exist. I don't want to go as far as that; I think that they do exist, but they are simply artifacts of our brain-processes. I would explain the quale of redness, for example, as a particular sort of activity in the brain that is triggered by light of a particular frequency arriving at the retina and so stimulating certain parts of the brain where light signals are processed (the visual cortex). The particular nature of redness has come about through the evolution of our brains: animals have had an advantage through associating certain colours with particular mental states (red standing for blood, for danger, and so on). Because of this each colour will trigger distinctive patterns of activity, qualitatively different in each case and associated with different sorts of emotions. What we call qualia are simply these evolutionarily determined patterns of neural activity. I agree with Descartes' picture in which everything that we call our world is just a certain pattern of nerve impulses, and the relation between these and external reality is indirect and theoretical. I'm convinced that we never actually perceive external reality; we only perceive our internal mental constructions.

MINNIE. Your view evades the main aim of science, which is to explain our world. The account of qualia that you have just given isn't an explanation at all, if you look at it closely. To begin with, we still have some way to go before we find a set of nerve impulses that might be a candidate for 'redness.' But let's suppose that research indicated that the account was at least factually correct, and that it could be shown that whenever someone reported a sensation of redness, there was always some distinctive pattern of nervous activity in the brain, a pattern that only appeared when redness was part of the person's awareness. Let's also suppose, for the sake of your argument, that this pattern had in it the sort of ingredients that one might expect on the evolutionary account, with a small excitation in the parts of the brain to do with anger, fear and so on. You are then claiming that this pattern simply is the quale red. So you have abolished the problem of explaining the quale by redefining the word 'red' to mean, not what I, or you, are experiencing when we report redness, but a certain pattern of nerve-impulses. You've actually bypassed the scientific goal of explanation by declaring that, because the nerve-impulses are always present whenever 'red' is present, then they are therefore the same thing. Even though they seem very different — nerve impulses are a theoretical construction detected with instruments, and located in the brain; redness is an immediate experience and located in the external object — yet you hold that, because one is never found without the other, they are 'really' the same.

FRODO. I can easily dispose of that objection. You are forgetting about the problem of different viewpoints, which our wise and learned Author *[you won't get round me that way — Au.]* called the fallacy of the shifting vantage point. When you study my nerve impulses, then you have to use instruments, and make theoretical deductions, and the nerve impulses are obviously located in my head. I have a different viewpoint: when I study my nerve impulses — or, to express it better, when I live my nerve impulses from the inside — it is going to be completely different. Then the nerve impulses are integrated into an internal model of the universe, in which they are linked with an internal representation of the curtain, or whatever the red thing is, and so when seen from the inside it appears as though the redness, and the curtain, are outside. This appearance has developed in us through evolution, so that we behave in a way appropriate to our environment.

MINNIE. Yes, I'll concede that experiencing from the outside and the inside will be very different things. But to explain our experience in these terms you are having to resort to Computerspeak, talking about 'models.' No doubt if pressed further you would start referring to 'programmes' in the brain corresponding to different patterns of behaviour. This sort of language is fine if you are reading a computer programme written by a human being, who has set it out in neat paragraphs (headed 'redness,' 'curtain' and so on). But when we look at the human brain, all we see is a seething mass of electrical activity. There are no neat 'models' sitting there; the models live in the theories of neurophysiologists, not in the brain. There is still no sign of how this mass of signals can divide itself up into separated aspects that we can call 'red curtain' and so on.

Why don't we simply take the obvious course, and say that when I perceive a red curtain, I perceive exactly that: the real object outside me, and not nerve impulses inside. Obviously there are all

sorts of nervous signal processing involved in attaching the words
'red' and 'curtain' to the experience; but why shouldn't the basic
awareness be exactly what it seems to be, thus solving all the problems without having to resort to vague and hypothetical computa-
tional structures in the brain that no one has ever observed?

FRODO. Now that really can't hold water for a moment, because our
senses can be deluded. Suppose I arrange a red spotlight on a white
curtain; then you might still report a red curtain, even though it was
nothing of the sort. Or, to take a different example, what about
after-images? If I look at a bright light and then at a white wall, I
might mistakenly report a dark blob on the wall. But I would be
totally mistaken. The dark blob clearly is something inside me, in
my retina, which I am mistakenly placing outside, because I have
evolved to place all signals that reach my retina 'outside.' This
seems conclusively to support the view I have been arguing for.

MINNIE. I can't deny that this 'projection,' of something inside so as
to appear outside, happens with after-images. But you are making
an exceptional flaw in the visual system serve as the pattern for the
whole system. The normal functioning of the system involves perceiving real things really outside. If it wasn't for this we wouldn't
have any concept of outside at all. Then when something goes
wrong, through over-exposure of the retina to light, the error is
fitted into the pattern of real outside experiences as best the brain
can manage. The fact that there are occasional exceptions to the
process of really perceiving real things doesn't invalidate the basic
idea.

4. The role of consciousness

Our two speakers are clearly not going to agree for some time.
Fortunately, we will later on be able to provide them with some more
information to help the discussion along, when we bring quantum
mechanics into the picture. But before that, we need to come back to
the concept at the heart of this chapter. So far we have been taking a

very minimal definition of consciousness as the fundamental givenness of the world to awareness. Many people would say that the solution to our problems requires us to take a much more concrete and specific definition; that we need a notion of consciousness that really 'does something.'

What is it that distinguishes a computer from a person? Consciousness. What is it that distinguishes a quale from a random pattern of nerve impulses? Consciousness. It offers the hope of explaining everything. Yet, unless we can say what it is, it will explain nothing. There is a danger that it will be like Descartes' soul: something postulated when one cannot think of any other explanation, but incapable of further analysis and acting just as a prohibition on further scientific investigation.

We need to make sure we know how we are using the word. By consciousness many people mean 'awareness of self.' But this begs the question, since we first have to decide what 'awareness' is. In fact, what I shall mean by 'consciousness' is simply awareness. So I am not restricting the word to 'self-consciousness' or any other special form of awareness. If we can decide what awareness is, then we can easily say what is awareness of self and all the other sorts of special consciousness.

From this viewpoint, consciousness starts to look rather less like a cure for everything than it did at first. 'Has a computer consciousness?' means 'is it aware of anything?' This doesn't seem to get us anywhere, since we then need a Turing test for awareness. A computer might tell us that it is aware, but if it has been programmed by Frodo the physicalist, then 'it would say that, wouldn't it?' A robot equipped with sensors can respond to things around it, but does that mean it is aware of them? And saying that the difference between random impulses and qualia is one of awareness could just be a restatement of the difference between things as seen from the inside and things as seen from the outside.

Let us take a different line of approach. Daniel Dennett, the principal bogeyman of this chapter, has talked about consciousness in terms of language production. One gets the impression from his book that it is all about the process of putting into words what is going on in one's head; or rather, what is going on in one's head is a scattered and complicated process of putting things into words, which we call consciousness. This is inevitable because of his approach of hetero-phenomenology, described earlier. He does not allow our actual experience as evidence, and so he is forced to make do with the words we tell about our experience. Consequently, consciousness comes out as all about words.

No one who has taken the trouble to study their own awareness can be satisfied with such an approach. Unfortunately, surprisingly few writers about awareness seem to have studied their own awareness. If you have practised any sort of meditation then you will have a good idea of the nature and workings of your own mind, and will know that words are only a very small part of it. But if not, then it will help you to take the time to reflect on what you are aware of. The following little exercise might help.

Choose an object that you are going to study — a flower or a stone is good, but it should be something that is fairly neutral emotionally. Sit in a comfortable position, with your feet firmly planted on the ground and the object in front of you. Then make sure your body is fully relaxed: pay attention to your breathing until it is peaceful, and then check over the different parts of your body to release any tension there. In a relaxed state, gently look at your object for ten minutes or so. Don't stare at it, but simply be with it, passively. Don't speculate about it or try to describe it, but be empty and receptive to the object itself. If (or rather, when) you find that your mind has wandered, just bring it gently back to the object.

Most people, when they do this, easily realize that, although there are all sorts of words running through their minds while they look at the object, the words are separate from the awareness of the object. Words are a babbling brook, in the foreground or the background; but the awareness of the object itself has nothing to do with the words. Examining the experience more closely, many people realize that the focus of awareness, so to speak, can be anywhere between the object and the self. Sometimes there is only the object, and the self might as well not exist. At other times one is absorbed in oneself and the object is very much out there, being viewed as if from a distance.

Equally, however, the object is inseparable from the concept of the object, unless one reaches quite a deep contemplative state. It is impossible, without much practice, to view a flower as merely an agglomerate of coloured surfaces; it is conceptually a flower, even if the word 'flower' is not explicitly running through the mind. So the result of the exercise is, for most people, the conclusion that perception is not a purely verbal matter — there is actually an entity independent of the conscious verbal processes that accompany it — but this entity is not independent of all mental processes. (I am all the time talking here about direct experience, not about the theoretical construction of hypothetical mental processes that might underlie this experience.)

Most importantly, the conscious perception of the flower is, for most of us most of the time, neither rooted purely in the internal conscious mind (as, perhaps, Descartes would have it), nor existing totally independently of our minds (as the naïve realist before Descartes might have imagined). Neither extreme actually reflects our experience, in which there is a mobile focus of awareness involving to different degrees both the external and the internal. Consequently, the conscious perception belongs neither to the self, nor to the external world independently of self, but to the relationship between them.

It is striking that this is precisely the conclusion that emerged from the growth of the quantum mechanical view of the universe that I described in Chapter 4. There physicists realized that it was impossible to regard reality as existing independently of the observer. The concrete happenings that form the stuff of physics emerge from the relationship, the dialogue, between the observer and the world. Whether we proceed from an analysis of our meditative experience, or we proceed from an analysis of the very specialized experiences — experiments — contrived by physicists, we reach the same conclusion that reality is relationship.

5. The world and the self

Our treading of this path arouses hares at every turn (if not red herrings). I have spoken above of 'the self' and 'the world,' and many will seize on this either as a cop out, or as the key to all mysteries. For some, consciousness is essentially something to do with the self, and the self is either thought of as something like Descartes' soul, or as some particular structure in the brain, as a model of the person. So our quest for consciousness becomes displaced onto a quest for the self, which seems just as difficult and elusive. So I need to make it clear that this is not what I ultimately want to say; only the limitations of the English language make the true concept difficult to grasp.

When we meditate on a flower, and observe a variety of types of focus in the experience, as I described earlier, we can analyse that variety into a spectrum extending between two poles, which we name the self and the flower. But we must be careful to realize that the experience comes first and the analysis second; the 'pure' flower and the 'pure' self (though neither can normally be experienced in isolation) are things that we abstract from the experience. So when I say that experience involves relationship, it is not that the self and the world are given first, and they then enter in relationship; rather, the experience precedes the terms which it is relating. We are given a web of relationships within which things appear as less than real secondary constructions.

Now normally you can't talk about a relationship without having the things to relate; relationship is always relationship *between* two previously real things, the terms of the relationship. Here, however, the order is reversed, and our normal habits of language become misleading. But while ordinary language leads us astray, the mathematician has for a long time been using an invented language that takes things in the correct order, a language called *category theory*. In this, the primary things are relationships, and things emerge either as special relationships that 'don't go anywhere,' that are bound up in themselves; or as aspects that certain relationships have in common. We are looking towards a similar approach being required not just in mathematics, but in all areas of life.

Because of this, when I wrote above of 'the self' and 'the world,' I did not want to imply that these were real things apart from each other. By 'the self' I was not meaning an independent metaphysical entity, but the much more superficial 'me' that we know through casual experiences of the sort that I have been describing. This superficial self is a construction of relationships, a web that is seen to dissolve as soon as one turns upon oneself the sort of meditation that

I started, inadequately, to describe above. And the world too, at this superficial level, is a construction. It is as wrong to place all one's faith in the unreal superficial self, as it is to place it in the unreal 'external world' hypothesized by science. That scientific external world is the most powerful construction that the human race has achieved; but it remains quite inadequate as the sole basis for the rich tissue of qualia that makes up our experienced world.

We cannot leave the issue here, however. For behind the superficial self of casual experience there are successive layers of deeper structures that can be called the self, with various sizes of capital S. Psychologically it could be claimed, with some justification, that any account of consciousness that did not take this into account was fundamentally deficient. The passage from infant to adult is marked by the successive development of these structures. Interruption of their development leads to various impairments of consciousness, and the essence of consciousness is inconceivable without the distinction between 'me' and 'not me' that the infant forms in its first relations with the mother. Surely this must lie at the very heart of any account of consciousness?

This takes us into a new area, for the self, in this sense, is very largely *unconscious*. In psychological terms, the self is what characterizes our identity as a person; it contains all the beliefs, instincts, attitudes and abilities that govern how we react to circumstances — all factors that lie below the surface of our normal conscious awareness. It is profitable to regard our behaviour and feelings as being governed by more than one layer of selfhood, from the façade that we adopt to impress colleagues (and to bolster our own self-esteem) to the deeper levels that are less changed by circumstances. Some would say that there is always a Real Self, an immutable core underlying all the layers of temporary masks; others would argue that the Real Self is something grown through experience and, very often, suffering. It cannot be presumed to be there in all. This view has been eloquently developed by Needleman (1993) and the role of suffering endorsed from a different perspective by Miller (1974).

If we examine our experience, it seems clear that there can be awareness without these deeper levels of selfhood. It makes sense to try to understand the simplest features of awareness — colour, sound, movement — without the deeper sense of the self. In this way, consciousness in a bare, primitive form comes before the self. But the self, in the deep sense, is what then makes consciousness *human* consciousness. It informs and structures the world that emerges from our more superficial awareness. At the same time, the self is formed and moulded by the experiences that impinge on us: at this deeper level, where we have gone beyond conscious awareness to bring in the unconscious elements that make us who we are, we find the same network of relationships, the same intertwining of external and internal, that characterized our superficial awareness.

From this point of view, 'consciousness' is too limited a word, because we need to bring in the unconscious to an even greater extent than the conscious. Yet the division between the two is itself shifting and uncertain. We can learn to become more or less aware of our deeper motives and impulses, so that there is a sense in which we are dealing with a broad field of experience, part of which is accessible to consciousness in the narrow sense, but all of which is essential within the interplay that constitutes our human world. In the next chapter I shall describe one model for trying to understand how these unconscious layers of the self play a part in constituting reality.

Chapter 6
A Quantum World?

We have seen how consciousness (or awareness) was essential to understanding reality, but that the question could be approached from two directions. Either one started from the totality of our awareness, the world as it appears to us, and then tried to make sense of it by the application of various world-models; or one started from a broadly scientific approach, recognizing that the extra dimension of consciousness needed to be included if one was to make sense of perceived qualia. In this chapter we shall see how far it is possible to go on this second path. In doing this, we shall be able to pull together many of the strands of the argument so far, using the underlying ideas of quantum theory in order to give a concept of reality that is more flexible than the naïve picture of classical physics, without being purely individualistic.

Most would agree that there does exist a world external to them (it is not completely a dream) but that a lot of what we call 'the world' is a construction — individual, joint or societal, built up in the way that I analyse later (see p.153). The paradigms of classical physics describe a completely definite world that is passively observed, while those of quantum theory describe a world of specific potentiality which becomes actual through a dialogue between the observer and the observed.

It is important to be clear about the contrast here between the quantum view and the view of Descartes, the originator of classical physics, which we examined in Chapter 2 (see p.51), Descartes had an observer, the immaterial soul, and an observed, the material universe, and out of their interaction emerged our human consciousness of the world, a consciousness that was, however, illusory in many of its most significant features. For Descartes both the soul and the scientifically understood external world were real, without any qualification; only our perceptions were subject to illusion.

What alternatives are there to the view of Descartes, and what do they tell us about the nature of reality? We look in turn at some of the more promising ways in which different authors describe the emergence of consciousness from a unified picture of the universe — a monistic view which does not introduce a soul as an entity of a quite separate kind from matter. This is not to say that one cannot talk about entities that were traditionally regarded as 'spiritual': Zohar (1991), for instance, is happy to use ideas like 'the self' and 'god,' but sees them as arising from a unified view of physics, not as something completely 'other.'

There are two broad types of approach. In the first (quantum theories of consciousness) we start from some version of quantum theory and then use it to explain the nature of consciousness. In the other (consciousness theories of the quantum), consciousness itself is taken as basic, and this is used to explain some of the mysteries that otherwise surround quantum theory. I will confine myself to two examples of each type.

1. Quantum theories of consciousness

In *The Emperor's New Mind* (1989), Roger Penrose puts forward a highly ramified argument to support the view that consciousness is in fact a consequence of a basic physical theory, at present only tentatively grasped, which unites quantum theory with gravitation theory — a new theory that he called Correct Quantum Gravity. Strictly speaking, he focuses not so much on consciousness, but on intelligence; it is, however, implicit in the argument that consciousness is an essential part of human intelligence, and that it is mainly through consciousness that Correct Quantum Gravity operates.

The key aspect to understanding human intelligence, according to Penrose, is problem-solving. And, being a mathematician, he takes as an example of problem-solving the sort of problems that are solved by mathematicians. It so happens that these, he claims, give evidence that something very special is going on in the human brain. He argues that there are some sorts of mathematical problem-solving that proceed in ways that no computer could emulate. He calls these non-algorithmic:

an algorithm is a fixed set of procedures followed mechanically, algorithms underlying all computer operations apart from those involving
some random input. Non-algorithmic problem solving, by contrast, is
a procedure where the correct answer emerges through an intuitive
leap that takes one outside any single fixed system of algorithms.

While this conclusion is probably correct, the argument that brings
Penrose to the conclusion is the Gödel theorem argument that we
examined earlier (see p.111) and showed to be incorrect. But, even
though the Gödel argument is wrong, the evidence does suggest that
human beings often solve problems by leaps of intuition which we
cannot explain in terms of mechanical processes. Let us suppose, for
the sake of argument, that there really is something going on here that
a computer — that is, a machine working on fixed algorithms —
could not emulate. This means that the physics of the human brain is
non-algorithmic; that the brain cannot be explained by classical
physics. (This final step, based on the idea that classical physics is
algorithmic, involves further technicalities that are still controversial,
which we will not go into here.)

Now there is one obvious candidate for something non-algorithmic
in physics: the collapse of the wave-function in quantum theory. The
transition from a superposition of possible outcomes to one particular
outcome is something that cannot be determined by any sort of
mechanical rule. So the interesting hypothesis emerges that the brain
uses quantum collapse to produce non-algorithmic solutions to problems. If the brain is thought of as a computer, then it has to be what
Deutsch has called a quantum computer: one that uses the non-
classical aspects of quantum theory explicitly in its working (and not
just as a means of implementing classical functions like switches on
microchips, as happens in conventional computers).

The final step in Penrose's argument, which will not concern us
here, is to use cosmological considerations (which are very controversial) to link quantum collapse with gravity.

For Penrose, reality is very much the scientific reality of the
Newtonian heritage. He does not follow up the indications given by
quantum theory that our conception of reality needs to be altered, but
instead regards the quantum state, scientifically described, as being
itself reality. There is for him no distinction between Bohm's explicate

and implicate orders. In terms of physics it is a defensible position, but it leaves untouched the questions about the true nature of consciousness, and in particular the understanding of qualia, that we described earlier. Consciousness becomes linked to gross gravitational changes in the brain, in a way that seems to explain nothing about its inner nature.

My second example of a quantum theory of consciousness is that of Iain Marshall, who starts from the nature of consciousness and, in particular, from the idea that it is somehow unified. This unity is easily overstated: careful analysis through trained introspection and meditation (particularly some forms of Buddhist meditation in which one observes the passage of one's thoughts in a detached way) suggests that the apparent unity of consciousness is illusory, and that there are many different strands of mind, fading off into comparative unconsciousness. But, as a first approximation, consciousness certainly has unifying aspects. There is a sense in which my many different perceptions and thoughts are given some kind of interrelatedness that enables it to be said that they are 'mine,' and that this interrelatedness is itself a quale, a primary aspect of consciousness, that cannot be reduced to structural aspects of the perceptions themselves. When I hear the rain outside and simultaneously see my computer screen, there is nothing internal to these perceptions that links them; they are linked only by virtue of their being my simultaneous percepts. So consciousness needs to be something that can link percepts that are associated with widely separated areas of the brain, such as hearing and vision. The unity of consciousness requires something unifying in the brain. It could all be unified by being brought to a single place in the brain, by nerve-tracts linking this 'seat of consciousness' with all the other areas. Many neurophysiologists have looked for something of this kind, and have put forward many different proposals for where the seat of consciousness might be, but there seems to be no physiological evidence for anything like this happening in the brain.

An alternative is that there might be a global quantum state in the brain which taps into all the different nerve processes. Normally physicists think of quantum states as referring to very tiny objects, such as atoms or electrons. But, as we discussed earlier, there is nothing to stop the application of the idea of a quantum state being applied to any system, however large. In some cases the physics of the situation forces one to think of a quantum state associated with a large system: this happens when the individual quantum states of the atoms or particles are linked together (much as the states of the two photons in the Aspect experiment are linked together; see p.89). In particular, at very low temperatures the quantum states can become linked in this way, resulting in phenomena like superconductivity, where the electrical resistance of certain sorts of materials drops to zero. The human brain, however, is not at a low temperature, and it is not made of super conducting material. Can similar things happen in it, despite this?

Marshall has proposed a theory of this kind, based on a mechanism proposed earlier by Fröhlich. For the process to work, we need to have a number of ingredients in place. First, we need a collection of oscillating molecules that are somehow linked together by forces between them. This could be provided in the brain by the molecules making up the membranes surrounding nerve cells, which are arranged in parallel and which are linked together by electrical forces between them. Linked oscillating molecules can exhibit wave motions that move across the whole array. Each such wave is associated with a quantum state, and the smallest quantum of such a wave is called a *phonon*. So far this is just a quantum description of something that could equally well be described classically: it is the same as happens when any solid vibrates to give off sound, and does not involve any special quantum effects. To get these, one needs a second ingredient, and a special technical condition.

The second ingredient is some way of supplying energy to the molecules to feed their oscillations (like pushing a swing to keep it going). Physicists call this 'pumping' the oscillations. In the brain this could be provided, in principle, by the chemical processes that fuel the propagation of nerve impulses. These processes involve the transfer of energy to protein molecules embedded in the cell membranes, and there is no reason why some of this energy should not go into the membrane oscillations. The technical condition concerns the frequencies of the phonon vibrations that the system allows. If we think of this as being like a musical scale, we need to have the intervals of the scale very widely spaced at the lowest pitches, and more closely spaced further up. This happens naturally in the case of, for instance, the various harmonics of a vibrating string or a bell; but, unfortunately, it does not seem to happen for the oscillations of the cell membranes, so that this could be an obstacle for the theory. There is a third ingredient, which in practice is always present, but which plays a special role in the mechanism: there needs to be a reservoir (such as the fluid bathing the nerve-cells) to take away the energy that is being pumped into the oscillators, so that the energy of the system remains in balance.

Once these ingredients are all in place, then it is possible in many cases to achieve what is called a 'Bose condensation of phonons.' Without quantum theory (that is, classically) the energy being supplied to the system is distributed uniformly over all the different vibrations, so that the net effect is that different molecules, unless they are near neighbours, will be moving in quite unconnected ways. Quantum theory, however, gives the possibility for all the energy being channelled into the wave of the lowest frequency, in which all the molecules in the entire system are vibrating in synchronization. This is the Bose condensation: all the phonons (waves) that make up the motion 'condense' together in the bottom frequency.

If something like this happened, then the coordinated vibration, or rather, its quantum state, would be the single entity that interlinked the different nervous processes going on in the brain. It would interlink them, because the global oscillation would be affected by all of them; everything that happened in a part of the brain that was participating in the Bose condensation would modulate its frequency, just as the many elements of a TV picture are brought together to modulate the single vibrational frequency making up a particular TV channel.

Marshall's theory, unlike that of Penrose, does engage with the details of what consciousness is and what is going on in the brain, though it still does not address the problem of qualia, and it still contains some speculative elements, though fewer than Penrose's theory. Both theories are open to experimental test: tests of Penrose's theory have been under way for over a year, conducted by myself and Dr Chris Nunn at Southampton (Nunn, Blott and Clarke 1994) and there is a large background of experimental work relating to the

Fröhlich mechanism in biological systems. It is worth noting that Marshall's theory is also important as an example of a very widespread way of thinking about living systems as being coherent, with the states of all their molecules locked together. These ideas have been recently discussed by Mae-Wan Ho who emphasizes that 'a coherent system is totally transparent to itself as all parts of the system are in complete, instantaneous communication. So in a precise sense, it knows itself completely.' (1993) In this way the unity of consciousness can be seen as a special case of an internal unity of being in all organisms.

If these theories are on the right lines then consciousness is at its base a quantum mechanical phenomenon. It means that our awareness has the capacity to implement the peculiarly quantum effects that we reviewed in Chapter 4: the capacity to be connected in its inner being with other quantum systems separated in space; the capacity to bring together into a superposition differing states of existence, then spontaneously to resolve into some definite state, and so on. A whole new range of possible ways of thinking about consciousness is opened up.

The drawback of these theories, however, is their controversial nature. The evidence that consciousness is really based on these very specific physical foundations is slender in the extreme, and the conservatism of many scientists will hold them back from proceeding down a path which promises much in terms of global understanding, but little in terms of explaining the current problems of how the mind functions. It must be stressed, therefore, that all the ideas we are now exploring do not offer the solutions to current research problems. Rather, they open up new ways of thinking that can illuminate, as we shall see later, deeper problems such as the nature of qualia, and which define wholly new research programs.

2. Consciousness theories of the quantum

We now examine examples of theories that start from consciousness as the fundamental ingredient and use this to shed light on quantum theory.

In Chapter 4 (see p.100), I described Lockwood's theory in which the quantum wave-function of the universe contained all the different possible outcomes of quantum choices, and particular outcomes depended on the way in which consciousness provided a particular perspective on the wave function, in which just one of the outcomes entered awareness. The idea brings in several new aspects not exhibited in the previous theories.

First, it takes seriously the fact, noted above, that the unity of consciousness is only a qualified or apparent unity. Lockwood draws ideas here from experiments on patients who have had the two hemispheres of their brain separated as a cure for epilepsy, who appear completely normal, but who, under special experimental circumstances, can exhibit what appears to be two separate and conflicting types of conscious behaviour in one body. Lockwood suggests that this may be an extreme manifestation of a normal state; that we should not think of consciousness as a single entity labelled 'ME,' but rather as a continuous field of perceptions and thoughts, with 'nearby' ones linked into a certain unity, but 'distant' ones going their independent ways. The different perspectives on the quantum wave-function is then just a generalization of something that is a fundamental part of the nature of consciousness.

Second, the theory takes very seriously, as does Penrose's theory, the problems that beset quantum theory when one tries to extend it from the realm of atoms to that of people and planets. Whereas Penrose looks for a physical solution that will then shed light on consciousness, Lockwood challenges us to examine the nature of consciousness itself in order to find a solution. Unfortunately, he gives us very little help in rising to this challenge. The physics of consciousness, in his theory, is only given a shadowy sketch, from which it is very hard to see how to proceed.

Evan Walker's approach (1970) came many years before those above, but it has the same general features. For Walker, consciousness is real but not physical: he recognizes that it is essential to include consciousness in our description of the world, but he claims that it is not measurable in an objective way. It is influenced by the physical world, but it does not itself influence the physical world. This one-way link between physics and consciousness passes through a quantum mechanical process connected with the operation of the nervous system. Like Marshall, he looks for global quantum states covering the entire brain for this physical basis for consciousness, but rather than choosing a global oscillation, he chooses a global electronic state, postulating that some electrons can roam freely through the brain by a process of 'tunnelling' from one molecule to another. (Tunnelling is a peculiarly quantum mechanical effect in which a particle — usually an electron — passes through a thin insulating layer that according to classical physics should prevent its passage.)

This use of tunnelling electrons is linked with his theory (1977) of the way in which nerve cells transmit impulses from one to another. The end of one nerve is separated from the dendrite (or body) of the next nerve cell by a narrow gap called the synapse. In higher animals the signal passes from one to another by means of special chemicals called neurotransmitters that diffuse across the synapse from one cell to the next. (The pharmacology of the brain is very much based on manipulating these neurotransmitters.) According to Walker, the trigger that causes neurotransmitters to be released is an electron that tunnels its way across the synapse from the second nerve cell to the first, in response to an electrical signal on the first nerve cell. In this way quantum mechanics lies at the heart of all nerve signals, and the collapse of the wave-function of tunnelling electrons therefore plays a vital part in thought. It is this that consciousness links into.

3. Narrow and broad quantum theory

In order to understand the quantum world of Zohar and others we have to go beyond the picture of quantum theory that was explained in Chapter 4 to a different way of looking at the theory. It is based on a formalism called 'quantum logic,' but extended to something considerably broader than just logic, which I shall here call 'broad quantum theory.' Endless misunderstanding has been caused by a failure to distinguish between this and the 'narrow' quantum theory that is the subject of most popular descriptions, and on which I have mainly concentrated here.

Earlier we focused on the idea of a dialogue with the universe, where questions were asked to a system in a certain state, and the system responded in a way that depended crucially on the question asked. We had here the two main ingredients of quantum theory: questions and the state. The many different interpretations that we examined in Chapter 4 (see p.97) differed in the way in which they interpreted the state. For some it was a real aspect of the universe at a given time; for others it described only a statistical situation holding over a large number of repeated experiments; for others again it described a kind of meta-universe, containing particular actual universes as many different worlds.

Broad quantum theory would hold that all this emphasis on the state puts the cart before the horse. The important thing in broad quantum theory is not the state but the questions. There are different levels of looking at the questions. The simplest level is one where all the questions are of the Yes/No form. Ultimately any finite piece of information can be reduced to the answer to a series of such questions, although in some cases one may have to pose a great many questions to narrow things down enough. (Twenty such questions, as in the usual form of this guessing game, allow you to distinguish between over one million possibilities, which just about covers an English vocabulary). A system of handling Yes/No questions about quantum systems is called *quantum logic*. The logic has to contain rules about which questions one is, and is not, allowed to ask at the same time. (We have already learned that, for example, one cannot ask an electron 'are you a wave?' and 'are you a particle?' at the same time.)

As a physicist, one may want to extend the sort of questions that one might imagine to questions like 'where *exactly* is the particle?' Of course, in experimental physics nothing is exact, but for the theorist it is useful to be able to imagine a situation in which ideally one could ask such questions. This takes one beyond Yes and No questions; however many of these you ask, you can never hope to specify a position with complete accuracy. Questions of this more general form are called *observables* — rather an ironic name, since they can never be observed in practice. Yet observables can still be thought of as ideal limits of endless sequences of Yes/No questions, getting closer and closer to precision.

Finally, within some of the more exotic theories of physics, one finds examples of 'questions' that one can ask of the mathematics, but which can never be asked in physical reality, even as ideal limits. The 'questions' share the same mathematical structure as observables and Yes/No questions (in technical jargon, they are all *operators)* but they correspond to nothing in reality. These last two levels of question are handled by a system more general than a logic, a system called an *operator algebra.* Like a logic, it contains rules for when questions can and cannot be asked at the same time.

We are trying to reverse the order of the cart and the horse, starting now with the questions. So how do we proceed from the questions to the state? This is the key point about broad quantum theory. The answer is a purely mathematical one: *the state is a numerical function on the logic (or operator algebra).* This means that the state is something that assigns a number (a continuously varying number — fractions and numbers like *pi* are included) to each question, this number being the average answer that would be obtained to the question. Thus, if the question is 'what is your position?' — understood as being measured against a ruler — then the number is the average position that would be found if you asked the question many times. In the case of Yes/No questions, the number is obtained by scoring 1 for Yes and 0 for No and then averaging the results.

For those unused to the ways of mathematicians and physicists, this sounds strange. Surely a 'state' should actually make some statement about the way things are? Yet here a state is merely a kind of statistic about answers to questions. What is happening here is that this new formulation, which does not change the content of quantum theory but only restates it with a different emphasis, is revealing the true nature of the quantum state. The 'state' does *not* say anything about how things are in some absolute reality. Rather, the state is no more than a specification of what we can come to know via a particular set of questions.

This has two important consequences. First, it offers a way out of the paradoxes surrounding the 'collapse of the wave function.' For on the broad view, whenever there is a change in the questions that one is asking — whenever, for example, one obtains new information or starts relating to the universe in a different way — then the definition of the state changes; one switches to a new ball-park with a quite different set of states, reflecting not a 'collapse' in some absolutely real universe, but simply a change in the questions being asked.

This solution to the collapse problem is similar to that described earlier (see p.98), which I described as a cop-out. There is, however, an important practical difference that emerges from the way broad quantum theory is actually carried out. Because of the emphasis on the algebra of observables, what happens is that this algebra itself becomes a very powerful specification of 'the way things are,' in a way that does not discriminate between the observer and the thing observed. Reality is thus placed in the nature of the relationship between the observer and the observed, precisely as required by the arguments we have so far conducted. The most fruitful applications of the method involve the construction of a network of algebras covering the innumerable relationships that spread across the universe, meshing together in a consistent way, but without there necessarily being any ultimate observables (the God's eye view) to define an ultimate reality.

The second consequence is that broad quantum theory is far more general than a quantum theory based on states — hence the name I have given it. If one starts with states then one is immediately thrust into the world of microscopic particles, wave functions and the rest. It is hard to see any relevance to the world of normal experience. But if one starts from a relationship, expressed through a set of questions, then the context is still completely open. We could be talking about particles. But we could be talking about tables, thoughts, people, decisions. ... Quantum theory is suddenly liberated from the physics laboratory to become a language that can be used to discuss the whole of our world. It is in this sense that Danah Zohar, whose views we consider below, uses quantum theory to illuminate the whole of human life.

This way of looking at things reverses the usual roles of classical and quantum mechanics. Usually we think of classical mechanics as 'normal' and quantum mechanics as a specialized, exotic subject only required in extreme situations. From the new viewpoint that we have reached, the 'normal' framework for debate becomes that of broad quantum theory, which simply sets out a range of possible questions that one might ask about the universe, while recognizing that some of these questions might be complementary, because they are framed from incompatible points of view (like waves and particles).

Classical theory is then the 'specialized' version of this, in which one makes the additional metaphysical assumptions that all the questions are going to be compatible, and that one can keep on asking questions until one reaches ultimate truth. When we drop these tendentious assumptions, we are left with the quantum world-view.

4. The quantum world-view

All the previous accounts have accepted a view of reality that follows the accepted scientific paradigm. None has taken up the view, which I have been building up from many sources including quantum theory, that reality is not something that is simply given 'out there,' as the scientific view would have it: rather, reality is a dynamic construction involving the person, the society, and the indefinite realm of the unknown that lies beyond what we have become aware of and made sense of. Out of this interaction comes not only the mathematical world of scientific atomism, but also, and more directly, the richly coloured world of our experience with all its irreducible qualia. Given that quantum theory has, in a large part, given rise to this view, can we not work towards a quantum mechanical conception of the world that does full justice to the new view of reality?

Danah Zohar, in her book *The Quantum Self* (1991), is perhaps the first to take quantum paradigms seriously as a way of looking at the world as a whole. Although she fully adopts Marshall's theory of consciousness, and might well not endorse as relative a view of reality as I am here putting forward, she opens up a vision of the world that has far more scope than that offered by the conventional scientific view.

The starting point is the appropriateness of the quantum language to human consciousness:

> If we reflect gently on the contents of our conscious minds
> at any moment, we are aware of a dim array of multiple
> thoughts, of 'possible thoughts.' These borderline areas of

consciousness, the 'twilight of the mind' spoken of by some poets, are most accessible just before falling asleep, in states of deep meditation or under the influence of certain drugs, but they are always there, on the edges of any act of concentration. Their reality is fuzzy and their future indeterminate, awaiting some act of realization. Without them there would be no basis for the multiplicity of poetic meanings that distinguish poetry from prose, nor food for fantasy and the imagination.

... In quantum terms ... this fuzzy, indeterminate margin of thought is the necessary precondition of all thought, reflecting the quantum origin of our thinking. It is the physical basis of our creativity and of our freedom.

Each act of concentration is an act of thought realization. Each of us has had the experience that the process of concentration collapses the wave function of a superimposed array of possible thoughts, though few of us might have expressed it this way before being introduced to a quantum vocabulary. By focusing on any one thought, that one becomes a classical reality and the others disappear like so many shadows in the night.

From this basis, she builds up, in the course of a fascinating account, a picture of the self as a quantum entity. The key aspects of quantum theory that she uses are the ability of quantum systems to unite into wholes; and the freedom (indeterminism) in quantum dynamics. It is worth looking at these two aspects in more detail.

Relational wholeness

The Aspect experiment (see p.89) showed how a system that seemed to be made up of two separate particles (photons) could in fact behave as if these were manifestations of a single system that responded as a whole to a dialogue with the observer(s). More generally, whenever 'two' systems (from a classical point of view) are in a state of interaction and are at the same time exercising quantum freedom, then they are no longer describable as two distinct systems.

Freedom

Physicists speak of quantum events as random, meaning that they are individually unpredictable although their statistics (average values etc.) can be determined. Although this can be attributed to 'hidden variables' — quantities that have a real existence but whose fluctuating values are unknown to us — such hidden variables would have to lie outside ordinary space in order to account for the results of the Aspect experiment; so most physicists prefer to regard the randomness as intrinsic, and not just a manifestation of our ignorance. Randomness is a negative term, expressing no more than our inability (to predict).When experienced from the inside, in the case of our own actions, randomness is called freedom.

While freedom is the absolute essence of what it means to be a self, it is relational wholeness that is decisive for the nature of reality and its relation to consciousness. Relational wholeness makes sense of the unity-in-multiplicity of consciousness; without it there would be nothing but a multiplicity of individual neurones, even individual atoms. It is thus the basis of the Bose condensation phenomenon on which Zohar's account is based. But it has much wider ramifications in terms of the relationship between the self and the world. If I am a quantum entity, then when I and another such interact, there is a sense in which, at some level, we are not two but one.

This idea, developed further by Zohar in *The Quantum Society,* leads us to the point where quantum theory, consciousness, and the nature of reality come together, because it can make sense, as nothing else can, of the qualia of perception, and in this way reinstate the possibility of a reality that depends as much on our awareness as on scientific theory.

In perception we observe something distinct from our body — a flower, a table or whatever. This observation is an act of entering into interaction with that object; of coupling together, through the medium of light, sound or touch, our body and the thing or person observed. Zohar has argued that we are essentially quantum entities, and we are permitted by physics to regard all objects, in particular objects that we observe, as quantum entities. But interacting quantum entities must be regarded as single systems: their states can no longer be split into separate states belonging to one or the other. Strictly speaking, when I observe something, I do not latch onto and interact with every part of it: my eyes do not penetrate to the core of every particle, but they engage only with certain aspects of the surface. So when I observe something, I form a whole with an aspect of that thing.

If the basis of consciousness is something like the system that Marshall describes, then when I talk about 'my consciousness,' all I mean is the inner nature of some quantum system, the nature of this system as experienced from the inside, rather than observed from the outside as happens in physics. And when I talk about forming a whole with an aspect of something external, I mean that this system, whose inner nature is my consciousness, is extended to include the external part. Thus my consciousness becomes the internal side of a union of part of me and part of the object; I am taking into myself, and experiencing from the inside, an aspect of the external world. In this way we at last achieve an explanation of the qualia, which seemed such an important part of consciousness but which were left out of account by the classical theories. The qualia are none other than the interior sides of objects in the external world. They are located in the external world, and not in our heads.

If this is true, then it completely overturns the whole of the scientific view of reality for the last three hundred years; for on the scientific view our perceptions are largely constructions of our minds, only remotely connected to the objects themselves. But on the quantum view our perceptions are genuine windows into the real inner nature of the objects. Quantum theory gives us back the world of colours, smells and sensations that the seventeenth century revolution drove us out of. Paradise has been regained!

It is vital, however, not to lose hold of all the other things we have learned about reality. We cannot think of this quantum awareness as simply passively grasping something that exists in a given external world that is quite independent of us. Along with sweeping away the scientific idea that we cannot genuinely know the external world, we also sweep away the idea that there is a completely independent world to be known. Our acts of perception can indeed be genuine acts of knowing, embracing an object that is external to us. But this act of knowing is a quantum event, and so the external object is to some extent created in the process of knowing it; its reality lies in the relation between us and the world, not in a world independent of ourselves. Moreover, the encounter between us and the world is usually heavily weighted in favour of ourselves, as we bring to bear on the encounter the whole weight of our past memories, our concepts and expectations, filtering the image through all the complex processing power of our brains; and all the world brings is the simple existence of the object before us. Inevitably, in such an encounter 99% of what emerges comes from within ourselves.

But though so much comes from us, and in particular from our memories and the patterns into which past experience has shaped our whole consciousness, yet the ultimate source of these memories and patterns is still the external world. And so our perceptions are amalgams of the internal being of the thing directly perceived, and the beings of countless past things perceived that have built up our patterns of consciousness. Zohar describes this patterning by past quantum interactions as 'quantum memory' — a concept for which we have no physical account at present, but which is clearly called for in any quantum account of perception.

This argument takes us beyond the physical into realms normally regarded as spiritual. For, if we examine our experience to try to find when we most vividly see the interior being of a thing, when we most seem to grasp the thing itself without distortion by our own preconceptions, then we find that it is when we are moved by love for the thing itself. More precisely, love is a condition that can either drag the object of attraction into ourselves, in a self-centred grasp that totally obliterates it with our own desires; or it can draw ourselves into the object, so that our own ego is obliterated by the object which we see in its own right. It is our own nature that determines which it is that takes place, just as, in Mahrer's account, the person determines whether she or he fabricates an external world 'out of whole cloth,' or accepts a ready-made external world. This activity is described by the Sufi mystic Ibn al-'Arabi (of whom more later) through the metaphor of the 'heart,' when he says that '... the heart encompasses the Reality'; and he continues, 'but though it be filled, it thirsts on.'

All these considerations apply not only to inanimate objects, but to other people. Zohar stresses that there are genuine acts of knowing another person, through partial unity of two quantum systems, which we experience in intimacy. These are vital in forming the self, which exists in relation to others.

Returning, then, to Zohar's account, we can start to understand how the combination of freedom and relational wholeness makes possible creativity, which is where part of the external world and the self are linked together in a common process of coming-into-being. Zohar uses the example of a child discovering how to make a clay pot:

> The whole process by which the child makes his pot is a chain of free decisions — the idea to make a pot in the first place (instead of a clay man or an aeroplane or whatever), then the idea to make this particular sort of pot, then the decision to press the clay a bit here and to round it off a bit there, etc.
>
> As these decisions unfold, the child slowly discovers both his pot and that within himself which likes to make such things, but this discovery is a creative discovery, because it is precisely through this unfolding discovery that the child makes both the pot and himself (his sense of beauty). His unfolding discovery literally snatches the pot and this aspect of himself from the shadowy realms of possibility and actualizes them. His creation has acted as midwife to the birth of a small bit of new reality.
>
> ... In following his natural quantum impulse to make an ordered relational whole from the data of his experience — his quantum impulse to integrate himself — the child *ipso facto* draws together an object (his clay pot) and a world (his relation to the clay pot, its meaning for him and for others) which were never there before. Child, object and world are all co-realized through the free and undetermined collapse of many possible children, objects and worlds in the child's mind. (1991, p.175f)

5. Quantum reality

We have now reached a conception of reality derived from physics that supports both the data of our personal consciousness and also the new outlook inaugurated by the postmodern philosophers. The formal structure of broad quantum theory has given us the technical tools for handling this new reality, while the vision of Zohar has fleshed out the formal structure so that we can understand how it applies to our everyday lives.

We find ourselves in a world where there is no longer a well defined notion of what is or is not the case. Instead we have a network of overlapping relationships and contexts. Within each context, there is still an openness; the questions that can be asked, forming a quantum logic, are not fixed, but are a matter of dialogue within the given context. The answers that arise change the context, and feed into other contexts and relationships.

Consciousness is one of these contexts, the one in which we have our immediate being. The older approaches would place the physical context first and try to derive the conscious one (see p.133) or try to add consciousness on to the physical context in order to complete it (see p.138). Now we see that we need to regard the conscious context and the physical context as each having their own validity and each determining the other in a dialogue, each falling within the overall descriptive language of broad quantum theory. The qualia of perception are the fruits of this dialogue, neither purely residing in the brain, nor purely in the thing observed, but standing as genuine acts of participation between the two.

We are now ready, in the next chapter, to take stock of the path that we have travelled, linking this quantum picture into the philosophical discussion, and then to explore what it means for our practical relation to the world that we call real.

Chapter 7
Alternative Realities

We have now reached the point where we begin to see the world as a web of interconnecting relationships, out of which emerge, more or less distinctly, ideas of self and not-self, objective and subjective. Science tries to make sense of this web in various ways which change from one period to another, as the prevailing paradigm changes.

In Chapter 1, I described how Kuhn drew attention to the way in which a change in paradigm affected one's entire world view, in two senses: the conceptual structure of the whole was changed, and new data appeared that was previously unnoticed — people suddenly 'saw' new things. Feyerabend argued that this shift had to be counter-intuitive: we needed to take alternative world-views from magic, myth and elsewhere to set alongside our existing view in order to make it possible to see new 'facts.' This means that actively exploring other ways of looking at the world becomes an essential part of science. The particular examples of space-time and quantum theory then showed how science contained within itself the possibility of radically different ways of structuring reality.

In Chapter 5 we also saw that the world view is a matter of consciousness as well as concepts, the two being inseparable, affecting each other. Changing world-view can be a matter of changing consciousness. We (educated Westerners) are all set in a particular sort of consciousness, produced by the society in which we are nurtured, influenced deeply by scientific ideas. For us there is a clear distinction between the 'objective' world, 'out there,' and the 'subjective' world, which is 'all in the mind.' From this standpoint, any deviation from this viewpoint (seeing angels, for example) is branded as insanity, and any argument that requires a different viewpoint is branded as superstition. The whole structure is an impregnable and self-contained fortress. But we must ask, is it defending society or imprisoning it?

If we start considering alternative meta-paradigms, alternative world-views, then it is natural to ask how far the process might go. Do we stop with the models opened up by quantum theory? Are there limits set by the nature of human biology? Is there some irreducible necessary structure, arising from the mere fact that we are human, which every world-view must share? Kant thought there was. He stressed that human beings were constructors of the world, necessarily working with sense data and never having contact with 'things in themselves'; but he also held that because of the basic nature of our senses and our minds, there were certain structures of experience and thought that must be common to all cultures, to all possible experience, calling such structures the 'transcendental *a priori.'* These structures could be taken as the basic starting point in any philosophy, as an absolute given.

But while his arguments sounded plausible, they were undermined by the fact that the particular fundamental structures that he came up with turned out to be nothing but products of his own era and culture: he identified the geometry of Euclidean space, for example, as such a structure; whereas today workers in relativity theory, including myself, routinely use and imagine other geometries in thinking about the distant parts of space.

I described how, following on from this, Tomlinson argued that a world-view, concepts of rationality, and concepts of truth were inevitably culture-dependent, and no universal criteria covering all possible world-views could be extracted, because any attempt to extract universals would itself have to use a system of logic and rationality that would in turn be culturally dependent: there is no God's eye view for us mortals. From a biological point of view, the human mind seems to be an open system, with no foreseeable limit to the ways in which it might adapt to different circumstances.

If this is true, then we should be cautious about stating what reality must be, what rationality must be, until we have looked at culturally different alternatives and compared them with our own. The problem is that we are liable to make the comparison using our own system of rationality; we have to step fully into the other culture before we are able to make a comparison. For most of us this is an impossibly large step to take. None the less, the effort is worth it. By setting another system alongside our own we start to open up the possibility of alternatives, and scope is made for a paradigm shift of the radical form described by Feyerabend.

1. A theoretical framework: Mahrer

How can we, as Westerners steeped in a certain scientific tradition, come to grips with this possibility of alternative paradigms? A possible theoretical framework has been provided by A.K. Mahrer, working from a psychological perspective. His is an early example of a humanistic psychology, which places the human person first, without assuming any particular given scientific world. He describes his position as follows:

> Human beings live in worlds of their own construction, endowing these worlds with specific functions, and constructing their worlds to serve those functions. Just as the individual person determines the nature of his effective environment, so collective persons determine the nature of their collective social environment. ... It is on this point that humanistic theory aligns with existentialism and parts company from the natural sciences, including contemporary psychology and psychiatry which accept the assumption of human beings and external worlds being separate and interacting.
>
> Yet it must be noted that our position is no more than an assumption. Kant and Heidegger made the person into the constructor of his world by arbitrary fiat, by adopting a particular assumption. Our discussion of the modes by which persons construct their worlds are exfoliations of this fundamental *fiat* — and not statements of fact about the way things really are. (1989, p.180f)

Mahrer then goes on to describe four different modes of construction of the external world, as follows:

1. Receiving an intrusive external world

In this mode I construct the external world by the manner in which I receive it. I play no determining role in creating the components of the external world which intrude themselves upon me. I neither create these components of the external world, nor do I select which components of the external world act upon me, nor do I determine that they will act upon me rather than some other object.

Most natural science psychologies adopt a theoretical framework in which the source of activation of external forces lies within the external forces themselves. ... Our model differs in two respects. First, this mode is only *one* of at least four modes of constructing an external world. Second, even within this mode, our system acknowledges the person's *active* role in using this particular mode. (p.185f)

2. Utilizing a ready-made external world

In the first mode of constructing the external world, the person is active in receiving that which is initiated in the external world. In the second mode, the person constructs the external world by a process of utilization (selection, capitalization) of what is ready-made. Although the person does not create anything new, the person constructs an external world by utilizing what is already lying about.

By using this mode, the person can construct his world through the ways in which he simply perceives what is there. ... Dirt is there in the room. I utilize what is already there by means of perceiving it. I have managed to construct a world of filth by means of sheer perceiving ... (p.192)

3. Conjoint construction of an external world

The conjoint construction of external worlds requires that two or more persons bring out what is potential in one another. You must seek something within me, and I must do the same with you. It is not right there, 'hanging out' for direct use as in the second mode. We must find it inside each other and evoke it. I fashion you into being some way, and you do the same to me. Existential phenomenology asserts that the person and the world (whether that world refers to another person or an object) are to be described as forever working upon each other. Rather than being acted upon, a person is right there, building the world which is in turn building the person. ... To describe a person as conjointly constructing a world with another person is to specify one particular mode by which 'man and the universe are universally interdependent and co-defining.'... In this process the other object need not be another person. The other object can be a natural thing: a bird, a pen, a mountain, a stream, a sidewalk, a bicycle, a tree. Interaction with the physical object may be such that the physical object is altered by virtue of our interactive responding to it, and thereby a new external world is created. Thus, in this mode of creating and constructing the external world, the person can interact with the objects of nature. (p.195f)

4. Fabrication of an external world
We now turn to the most active role that a person can play in constructing an external world, i.e., fabricating an external world out of whole cloth. In this mode, the person is the complete architect, designer, and builder. The fabrication of an external world is an act of personal artistry, calling upon true creative resources. It is the mode of creative inventiveness. When one fabricates an external world, the possibilities of what can be created exceed those of the first three modes. You can construct any kind of external world — from a new city to a new piece of art, from a political ideology to a God, from a vision of Christ to an image of your ancestor, from a voice calling to you out of the crowd to snakes surrounding your bed. All you require is your own personal resources. (p.211)
    ... the restless urge to fabricate extends well beyond the personal and mundane. Most of traditional research is merely evidence of a person (the research scientist)

fabricating an external world from building blocks of
external reality. This person, in using our contemporary
research methods, is as far away as possible from the naïve
receiver of the intrusive external world (the first mode of
constructing an external world). The researcher is an *active
fabricator* of the external world. (p.215)

In this framework we recognize many of the themes that have already
emerged: the selectivity of perception in mode 2 recalls Kuhn's
description of the way in which a theoretical structure conditions what
things are noticed and what are not (so that Novae, new stars, are not
noticed until Galileo opened up the idea that things in the sky could
change); or, in Feyerabend's terminology, what counts as a fact, as
data, itself depends on the theory in use.

The mutual interaction of person and object in mode 3 clearly
recalls the situation that we met with in the case of quantum mechanics, where the very act of observing a thing in some particular chosen
manner determined the state of the thing observed — not by physically interfering with it in a mechanical manner, but in the sense that
until the observation took place, that state was only one among many
potential states, and a particular state was brought into being by the
mutual interaction of person and object. That led us to describe reality,
in such cases, as lying in the interaction between the person and the
object, and not in either one separately.

The fourth mode, the most radical, recalls Feyerabend's account of
the creative process involved in the formulation of new theories, but
is also much more critical about reality, pointing forward to the ideas
about alternative realities and religious realities that I examine next.

Mahrer's approach seems strange to people, like myself, brought up
in the conventional scientific tradition, in which reality is scientific
reality, and everything else, from visions of Christ to snakes round the
bed, has to be explained in terms of a fixed scientific reality. And yet,
if we take seriously the messages coming from quantum theory and
from writers who look critically at what underlies scientific theories,
then we have to expect a much wider range of possibilities than was
ever thought possible. Here Mahrer's scheme could give us a compass
to chart our way across these unknown seas.

In the following sections, we will look at particular examples of alternative ways of building reality in the various modes described by Mahrer.

2. The !Kung

I start with an example that has received considerable attention from anthropologists, although many others could be taken.

In a famous article, R. Katz described the healing trance dances of the !Kung (indigenous people of the Kalahari region of Africa) as follows:

> The central event in the healing tradition is the all-night dance. Sometimes as often as four times a month, the women sit around the fire, singing and rhythmically clapping as the night falls, signalling the start of a healing dance. The entire camp participates as the men, sometimes joined by the women dance around the singers. As the dance intensifies, n|um ('energy') is activated in those who are healers, most of whom are among the dancing men. As n|um intensifies in the healers, they experience !kia ('a form of enhanced consciousness') during which they heal everyone at the dance. The dance usually ends before the sun rises the next morning. Those at the dance confront, celebrate and reaffirm the spiritual dimension of their daily lives. They find it exciting, joyful, and powerful. "Being at a dance makes our hearts happy," the !Kung say.
>
> While experiencing !kia, one can heal. Those who have learnt to !kia-heal and are said to possess n|um, are called n|um k||ausi ('masters of n|um' or simply 'healers'). n|um resides in the pit of the stomach and at the base of the spine. As the healer dances, becoming warm and sweating profusely, the n|um heats up, becomes a vapour, and rises up the spine. When it reaches the base of the skull, !kia results. Kinachau, an experienced healer, talks about the !kia experience.

> You dance, dance, dance, dance. Then $n|um$ lifts you up
> in your belly and lifts you in your back, and then you start
> to shiver. *[n|um]* makes you tremble, it's hot. ... Your eyes
> are open but you don't look around; you hold your eyes still
> and look straight ahead. But when you get into *!kia,* you're
> looking around because you see everything, because you see
> what's troubling everybody ... $n|um$ enters every part of
> your body right to the tip of your feet and even your hair.
>
> $n|um$ is held in awe, considered powerful and mysterious.
> It is this same $n|um$ that the healer 'puts into' people in
> attempting to cure them. So, once heated up, $n|um$ can both
> induce *!kia* and combat illness. (Katz 1981)

The reaction of most Westerners to accounts such as this is to try to fit it to their own paradigm. There are some physiological points of contact which may be transcultural — the burning of $n|um$ is reminiscent of descriptions of 'the burn' in aerobics and 'the pain barrier' in athletics — and these can be assimilated in terms of the reaction of the body to exercise, hyperventilation and so on. The healing process can perhaps be described in conventional psychological terms. Nothing can force us to come outside the fortress of our own world-view if we do not wish to.

The alternative is to consider the possibility of the viewpoint of the !Kung as an independent world-view, one which is certainly inadequate for doing modern science, but one which is probably indispensable if one wishes actually to take part in the dance. Once one lets go of the desire to integrate everything into the scientific world-view, then one starts to see links with a variety of alternative views, and with the alternative areas of experience, of consciousness, that go with such views. We are in fact dealing with one form of a universal human phenomenon: the entry into an alternate reality through religious trance.

The phrase 'alternate reality' was popularized by Castaneda (see p.165) and has recently been analysed in detail, in the context of religious trance, by Goodman (1988). There are frequently specific details in common between the !Kung experience and that in other societies. The trembling or shivering, for example, in the earlier stages clearly links with the shiver that is often seen in Voodoo when 'spirit possession' takes place, as well as in the onset of trance in Pentecostal and related religions (Goodman, 1988 p.61). So there is indication here of a distinct paradigm of consciousness-change, expressed either as a state *(!kia)* or as possession by an external agency, depending on the belief-system in force. Without adopting this paradigm, one cannot learn to have the experience.

The state of *!kia* clearly can be described as a different state of consciousness, in which one perceives a different world. The change in external perception is described in the words 'you see everything.' This links with the ability of *seeing* described by Castaneda, examined later, and with the accounts of many healers working in indigenous traditions, where the paradigm in which they operate has a natural place for an enhanced perception, only loosely describable as 'seeing,' in which one becomes aware of the significance of external things, or the state of wholeness of other people. This phenomenon is widespread in most ecstatic religions (Goodman 1988, p.44).

Further accounts of the experiences of the !Kung (Goodman 1988, p.75) show that it is not enough to describe the experience as one of 'seeing' the normal world through physiologically distorted 'vision': experiences that are described as one's spirit leaving the body and performing feats in distant places, or in places in a different world, show that the phrase 'alternate reality' is an appropriate description.

The scientific paradigm can assimilate these experiences to itself if it tries, by providing explanation of abnormal physiological states in which perception is distorted and hallucinations occur. But the scientific paradigm cannot help in producing and controlling these states, or in understanding their significance for the person and the society undergoing them. Other paradigms — we might call them semi-scientific — might fare better. We see one of these at work in the case of the !Kung when they describe *n|um* as the flow of a subtle fluid. Such pneumatic descriptions of the human body have never taken root in Western culture (with the exception of the flow of blood, which had quite different methodological origins) although Freud advocated them strongly. They were also proposed, in a form very similar to Eastern pneumatic models, by Freud's disciple (and later opponent) Wilhelm Reich on the basis of orgasmic experience. But while Freud's psychology had enormous influence, the physiological ideas that were linked with it were soon forgotten. As for Reich, his exclusion from the Freudian establishment drove him to increasingly extreme intellectual positions, and eventually to a complete paranoia which discredited his entire body of work. Only now is its influence starting to re-emerge, and practices such as Reichian massage result in experiences of a flow that does indeed go into 'even your hair.'

In the East, however, ideas such as n|um would be familiar. The rise of n|um up the spine is clearly reminiscent of the rise of the serpent in Kundalini Yoga, and of the flow of Chi (Ch'i) in Chinese physiological theory. Here again, moving into the paradigm is linked with changing consciousness and having the related experience. The Chinese system of balanced movements, linked to the breath, called Tai Ji Quan (T'ai Chi Ch'uan), which leads to a deep stillness of mind and body, involves an experiencing of the flow of Chi through the limbs.

Perhaps the most extensively developed of the concepts of alternative reality found among indigenous peoples like the !Kung is the concept of Dreaming among the Australian Aborigines. The English word 'dreaming' — the term favoured by the Aborigines themselves — translates a technical term for a whole separate order of reality that can be accessed by dreaming in the usual sense, but which is more often accessed by ritual, involving physical activity and mental focusing. The explicit theme of the ritual is usually one of mythological events, but understood in a particular sense. As the anthropologist Elkin puts it:

> History there is, but it is the myth of that which is 'behind'
> or 'within,' rather than 'before' the present. And that which
> is 'within' is the Dreaming, the unseen, but permanent
> reality of each and every creature and phenomenon and of
> their relationships. (1954)

He describes his observation of such rituals very much in terms of entering a separate world:

> In those rituals we were 'in the Dreaming.' We were not
> just commemorating or re-enacting the past. Whatever
> happened in the mythic past was happening now, and there
> is no doubt that the men were 'carried away' by the
> experience ...

Goodman (1988, p.73) also cites evidence that Aboriginal rituals use trance in just the same way as other cultures to enter the alternate reality.

In this case, although in one sense of the word there is a single alternative reality denoted by 'Dreaming,' in another sense each clan has particular responsibility for part of this reality, so that its members would speak of 'my dreaming,' a term that denotes both this part of reality and also the ritual used to access it. Indeed, it is very appropriate to speak of a whole alternate world, one linked geographically to the ordinary physical world, because no one tribe, much less one person, has access to the whole body of ritual and legend. Each person's dreaming is part of a much greater whole, and the dreamings of different tribes are linked by paths (Elkin 1954, p.175) or songlines (Chatwin 1988) that are both geographical and inter-tribal.

Many modern shamanistic teachers have taken up this concern with dreaming, in the usual sense of the term, as something that links in with the enlargement of consciousness which, for the shaman, is an experience of the body (for the shaman is free of that separation of body and mind that has, as we have seen, been characteristic of Western thinking). As Arnold Mindell puts it: 'Dreams are snapshots of body experiences that are trying to happen, and body experiences mirror dreams.' (1993)

Let me stress once again that I not maintaining that these phenomena are inexplicable on the Western paradigm, or that they provide evidence for paranormal cognition. Those are quite different sorts of arguments. The point that is emerging here is that in order to do certain things, to operate effectively in certain societies, it is almost indispensable to adopt an appropriate paradigm. The paradigm and the society are linked. And, since the paradigm then moulds the nature of the world, each society creates its own reality.

3. Crossing the boundary: Carlos Castaneda

If reality, as many argue, is conditioned by the culture in which one grows up, then it is hard to see how it can ever be possible to make sense of the reality of another culture. The best that an anthropologist observing the !Kung, for example, could do would be to try to give some kind of a translation of their reality in terms of the anthropologist's own reality; and such a translation, as with all translations, will miss out many essential components. Indeed, since we are dealing with the nature of different sorts of conscious awareness, then it could be that the essence of another culture's awareness of reality is absolutely untranslatable: we are dealing with different qualia, and, as we saw earlier, qualia are by definition irreducible, not capable of being translated one into another.

Traditional anthropology had grown up with the view that it was by necessity limited to this sort of observation purely from the outside. As a result, there was great excitement in the 1960s when a report appeared that a young anthropologist had been attempting to cross the boundary between one culture and another in order to experience at first hand the different reality associated with a different culture. Carlos Castaneda's first book, *The Teachings of Don Juan,* appeared to be a popularization of the research thesis that emerged from this experience, complete with an appendix similar to the notes for a Ph.D. thesis. It described how Castaneda, a young anthropology researcher, became apprenticed to a sorcerer of the Yaqui tribe of Mexican Indians, and learnt the use of hallucinogenic drugs in changing consciousness. Further books continued the account, with *A Separate Reality* describing a phase where the apprentice is trained to use these drugs in order to enter specific states of perceiving alternative realities. The techniques used involve 'stopping the internal dialogue' — the constant internal monologue that we deliver to ourselves, shaping and moulding our ideas all the time — and 'stopping the world' through attention to 'gaps.' This rather mysterious concept appears to be either a gap in time, a kind of extension of stopping the dialogue to the entire universe, or else an almost physical gap that appears as a door between the ordinary and the

alternate reality. A similar sort of physical gap is described by Goodman as:

> the crack between the earth and the sky on the horizon.
> According to a tradition known around the world since
> ancient times, that is the hole, a kind of slit, that leads from
> our ordinary realm into an alternate one. (1990)

The correct attention to gaps can then result in the ability to *see,*
defined as 'a special capacity ... which would allow one to apprehend
the "ultimate" nature of things ... an intuitive grasp of things, or the
capacity to understand something at once, or perhaps the ability to see
through human interactions and discover covert meanings and
motives.' (Castaneda 1975)

Later books then focus on ways of altering consciousness without
the use of drugs. *Journey to Ixtlan* develops the ideas of knowledge
and power, and the nature of the human person, as well as further
techniques, while the fourth book *Tales of Power* propounds a metaphysics for the whole world-view. Several more follow in the series.

As the books described increasingly strange experiences, the
question of their authenticity started to be raised. Had the success of
the first work gone to Castaneda's head, causing him to fabricate the
sequels? If the later books were fabrications, could one trust the first
one? Does the progression, from the drug-based practices of the first
books to the drug-free practices of the later ones, represent no more
than a response to market forces as the years progressed from the
drug-oriented Sixties to the more puritanical Seventies? The questions
were investigated by C. De Mille in *Castaneda's Journey* (1976), in
which he presented strong evidence that all the books were in fact first
person novels, based on anthropological and other literature rather
than on first-hand field experience in Mexico. This takes away some,
but by no means all, of their value. The close parallels between what
Castaneda describes and other areas of shamanistic experience show
that he is presenting genuine anthropological material, even though it
is at times distorted, and, as I shall describe shortly, the books bring
together and comment upon a wide range of genuine experiences of
alternative reality.

Parallels

The first comparison, as I have already noted, is with the way that the seeing in the early books links clearly with the !Kung account of the way in which the healer attains a state where he can 'see' what is wrong with people. In many ways the (perhaps fictional) account of Castaneda is reinforced by the anthropological work of Taussig (1986). He has studied the experiences of Colombian Indian shamans, who use a hallucinogenic drink called *yagé* prepared from a vine. One subject recounts that 'Well, then I *saw* — as the *yagé* healers say,' (p.453) and he elucidates this seeing by explaining that: 'I was in another room — the *yagé* making me see. ... The sick woman is there and I am there seeing the sickness and the pain that she had in her head ...'

The same stress on seeing comes at times in Goodman's work, for example:

> "There was also the sense of seeing, in that there was something I needed to see, but again in the wider sense ..."
> (1990, p.211)

There are also some parallels with the techniques of the blend of European magic and native American wisdom that has been described by Starhawk. Castaneda's activities take place within the real landscape of Mexico, while Starhawk's take place within the interior landscape of the psyche, which extends into the collective psyche which she calls the underworld (1988, p.53); but for both of them there are common concepts like the place of power that has to be found so as to provide enough security to venture on the journey. In general terms, the ability of *seeing* which the later, non-drug books are describing is reminiscent of Starhawk's definition of 'magic' as the art of altering consciousness at will. (1988, p.13) This echoes Dion Fortune's definition as 'the art of causing changes in consciousness in conformity with will' (in Burnett 1991, p.134.) Both share the same approach to reality, expressed in Starhawk's epigram: 'Consciousness determines reality. Reality determines consciousness.'

Further anthropological evidence comes from Goodman's later work

(1990) where she shows that this change of consciousness into the religious trance can reproducibly be stimulated in most people by simple techniques of rhythmic stimulation. In this case the alternate reality is further removed from the standard reality: it appears first as an internal, dreamlike state; but it does turn out to contact ordinary reality in quite precise ways. Particular geographical features of the ordinary landscape can have new aspects revealed in the alternate reality, for example.

It should be clear from these comparisons that there is ample corroborative evidence to show that Castaneda's stories are depictions of experiences, rather than mere romancing. As it was expressed by Nevill Drury:

> Carlos [Castaneda] himself is probably the actual visionary and many of the shamanistic perspectives have probably been implanted in the personage of the real, partially real, and unreal being known as don Juan. In this sense it hardly matters to the person interested in consciousness and states of perception whether don Juan is real or not since the fiction, if it is that, is authentic enough. (in Burnett 1991, p.188)

Alternate reality

What clearly emerges is the link between consciousness and world-view; and between consciousness and cognition *(seeing)*. If one could switch, at will, between different modes of perceiving things, so that in one mode one perceived a world full of 'ordinary people,' in another mode a world best described as full of luminous eggs with inter-linking tendrils (to cite one of Castaneda's examples) then it becomes problematic which is to be called the 'real' world, and which image the 'real' human being. It becomes reasonable to talk, as in the earlier books, of there being several different worlds, with the passage from one to another guarded by spiritual presences that could take perceptual form to the person who was suitably trained.

The variety of possible worlds can be made sense of in terms of

Mahrer's classification. In particular, many passages of Castaneda recall Mahrer's third mode of construction, in that they describe how don Juan and his pupil talk about the way the world is being perceived in such a way that an alternative way of perception emerges from their joint experience.

De Mille, in his critical book on Castaneda, firmly rejects the idea of a separate reality. He accepts that people have funny experiences, both with and without drugs, which may in some cases lead them to talk about separate realities, but he insists that throughout all these experiences there is always one 'Boss Reality,' which is the reality of our ordinary perception. The so-called different realities are, according to him, in no way equivalent: the Boss Reality is the solid ground on which we can live and understand what is going on, and the other 'realities' are to be thought of as different sorts of hallucinatory distortions of it. This view might be supported by the evidence from other cultures that we mentioned along with the !Kung: it seems as though there is an everyday reality, which seems much the same in most cultures, and another reality that is entered through trance and which is culture-dependent. This makes it very tempting to say that the everyday reality is the Boss Reality, the one that is normative for our description of the universe, and the others are subsidiary states induced by the culture.

But there are a number of arguments against this. The most important is the fact that, while trance seems to be required in order to enter the other reality fully, its presence, in the culture and in the person's experience, strongly conditions the nature of the everyday reality. It could indeed be said that this conditioning is the main aim of many religions, rather than the peak trance experiences. Thus division between Boss and subsidiary reality is no longer clear-cut. The fact that one reality seems to be more universal than the other is explained by the religions involved in terms of the way in which during religious trance we use physical forms and images to make sense of non-physical input, and the choice of these images is culture dependent, though the primary experience may be less so.

Certainly Goodman, on the basis of detailed research into the form of alternate reality contacted by her techniques, is clear that the two realities are complementary, with neither subservient to the other. Her alternate reality is the abode of Spirits, and she writes:

> [The Spirits] know something that we in the West all too often forget, namely, that the ordinary and the other reality belong together. They are two halves of one whole. Only their joining will make a complete world worth living in. The existence of humans is empty without the Spirits, but theirs is equally incomplete without involving us, and the world about us. (1990, p.55)

In some cases, indeed, the alternate reality — far from being, as De Mille claimed, a subservient reality — is experienced as more real than the ordinary reality. This constitutes a most significant objection to the Boss reality notion, to which we shall return in the last chapter (see p.188.) Although the trance experience is in some sense 'abnormal,' in requiring special mental activities and usually a small amount of training, its content is reported as being more 'truly real' than 'normal reality.'

Alternate and graded reality

The final argument against De Mille's dismissal of alternate reality in favour of the Boss reality is, in my mind, the most crucial. It is an argument that takes us into a different conceptual framework that is crucial from the point of view of actual practice.

So far I have talked about alternate realities as if they were quite distinct things, one requiring special shamanistic techniques, the other not requiring anything special except being brought up in a particular society with that society's vision of reality. It is clear, however, from Mahrer's depiction of the different ways of constructing reality that these are not quite separate states of mind, but are simply points within a continuum. We are very fond of categorizing, or arranging things in clear stages; but in nature things are usually much more muddled and complex than that. The qualified shaman is able to jump straight to a distant point on the spectrum; but in learning shamanism, the apprentice progresses through steadily deeper stages of trance and steadily more definite visions.

More important still, the different realities are not exclusive, so that entering one automatically shuts out the other. The classical Tungu shaman conducts his spirit journey while at the same time describing it to the onlookers, so that he is in both worlds at the same time (Goodman 1988.) Or, to describe the situation more accurately, his world is enlarged so that it includes the spirit world in addition to, not instead of, the ordinary world. This is expressed in many shamanistic cosmologies, where the universe is described as having three layers: the normal one, together with layers above and below it (see Burnett 1991), with clearly defined means of communication extending between them. We are thus not talking about a switch into a different reality, so much as an enlargement of vision so as to discern broader features of what may ultimately be a single reality.

As we shall see next, different religious approaches vary in the way in which they approach the relations between these different aspects of reality, with Western religions stressing an integration of the different types of reality into a single world that manifests both material and spiritual qualities.

4. The *nagual*

With Castaneda's later books comes a shift of emphasis. There is no longer so much talk of two or more different realities, with the possibility of moving between them. Instead, a distinction is made between reality itself, in the sense of the structured, conceptualized world of ordinary consciousness, the world inhabited by things to which we have names ('people,' 'tables' and so on); and the unknown area surrounding this, consisting of what has not got a name, what has not been conceptualized and accommodated to our world — what is not part of normal 'reality.' Castaneda calls the conceptualized world the *tonal,* while to the infinitely wider unconceptualized area he gives the name *nagual* (pronounced 'nar-hwal'; De Mille objects that this is a blundering misuse of an Indian word with a quite different meaning, but here I use the term with Castaneda's meaning.) Below is the account given of the exposition of this concept by the sorcerer don Juan in *Tales of Power.*

Castaneda's description of the theory of the *tonal* and the *nagual* is set in the form of a discourse delivered to him by his mentor, don Juan, while they are seated at a restaurant. As usual in his books, the discourse takes the form of a Socratic dialogue in which don Juan is in total control and Castaneda plays the role of the simpleton stooge who prompts for more explanation. Don Juan begins by cryptically describing the *tonal,* the organizing, naming, conceptualizing faculty that makes up the whole of our explicit awareness:

> "The *tonal* is the organizer of the world," he proceeded. "Perhaps the best way of describing its monumental work is to say that on its shoulders rests the task of setting the chaos of the world in order. It is not far-fetched to maintain, as sorcerers do, that everything we know and do as men is the work of the *tonal.* ... At this moment, for instance, what is engaged in trying to make sense out of our conversation is your *tonal;* without it there would be only weird sounds and grimaces and you wouldn't understand a thing of what I'm saying." (p.122)

From the point of view of the shaman, however, the *tonal* is ambiguous, because while it is essential to human life, it closes off access to the non-conceptualized part of the world (later called the *nagual)* which is essential to the shaman's power:

> "The tonal is, rightfully so, a protector, a guardian — a guardian that most of the time turns into a guard."
> I fumbled with my notebook. I was trying to pay attention to what he was saying. He laughed and mimicked my nervous movements ...
> "The *tonal* is everything we know," he repeated slowly. "And that includes not only us, as persons, but everything in our world. It can be said that the *tonal* is in everything that meets the eye." (pp.122–4)

He then goes on to confront the paradox that every conceptualization, in other words every 'thing' that is isolated as such, is a product of the *tonal,* and yet the *tonal* can produce nothing from scratch, always requiring input from outside itself.

> "The *tonal* makes the world only in a manner of speaking. It cannot create or change anything, and yet it makes the world because its function is to judge, and assess, and witness. I say that the *tonal* makes the world because it witnesses it and assesses it according to *tonal* rules. In a very strange manner the *tonal* is a creator that doesn't create a thing. In other words, the *tonal* makes up the rules by which it apprehends the world. So, in a manner of speaking, it creates the world." (p.125)

After this rather abstract exposition, don Juan moves on to a more concrete illustration, using the restaurant as a visual aid, introducing the concept of *nagual,* complementary to the *tonal:* the unconceptualized, unformed; the inexpressible:

> "The *tonal* is an island," he explained. "The best way of describing it is to say that the *tonal* is this." He ran his hand over the table top.
> "We can say that the *tonal* is like the top of this table. An island. And on this island we have everything. This island is, in fact, the world."
> "If the *tonal* is everything we know about ourselves and our world, what, then, is the *nagual?*"
> "The *nagual* is the part of us which we do not deal with at all."
> "I beg your pardon?"
> "The *nagual* is the part of us for which there is no description — no words, no names, no feelings, no knowledge." (p.125f)

Castaneda, in his stooge role, then goes on to name possible candidates of the *nagual,* ignoring don Juan's prescription that the *nagual* is precisely what cannot be named. To each named candidate — mind, God and so on — don Juan replies that it is part of the *tonal,* and represents it by a knife, a fork, a ketchup bottle ...:

> "If the *nagual* is not any of the things I have mentioned,"
> I said, "perhaps you can tell me about its location. Where is
> it?"
> Don Juan made a sweeping gesture and pointed to the
> area beyond the boundaries of the table. He swept his hand,
> as if he were cleaning an imaginary surface that went
> beyond the edges of the table.
> "The *nagual* is there," he said. "There, surrounding the
> island. The *nagual* is there, where power hovers."

The novel then becomes a dramatic portrayal of this distinction between the conceptualized and the unconceptualized. It is a philosophical point that takes on vital practical reality, however, for the practising shaman, whose entire work is concerned with drawing power from the dangerous area of the *nagual* in order to alter the *tonal.*

If the concept of several distinct realities was questionable, it seems that the idea of the *nagual* returns to firmer ground, both psychologically and philosophically. A recurrent theme so far has been the ambiguity of the term 'reality.' It is not to be identified with the world of immediate appearances, because these are shaped and moulded by our own ideas, memories and expectations. But even less is it to be identified with the theoretical world picture of science, an intellectual construction erected on top of our perceptions and thus at one further remove from whatever is 'out there,' if anything. We would like to say that reality is this 'out there,' and yet if we say that we seem to be saying nothing because we are talking about a hypothetical realm about which we can know nothing. And thus we seem led, by such arguments and by the evidence of quantum theory, to place reality at the interface between the unknown and the known; at the point of disclosure where what is other than ourselves enters into a relation with us that changes both parties to the exchange for ever — reality in relationship.

This encounter and relating with the unknown is something we do only at rare moments of life, when our world is enlarged by a quite new experience. We can only stand such revelations in small doses. 'Humankind cannot bear too much reality.' Our whole adult organism seems developed to keep us within the safe conceptualized real of the tonal. If I compare my reactions to those of a young child, for example, I realize how the child spontaneously notices a host of events in the environment that I now screen out; how the child is plunged into emotions of grief or joy that I now carefully hold at a distance, keeping on an even keel. Our emotions warn us when are stepping outside the safe confines of the tonal, and often bring us to heel.

Yet the boundaries of the known can fascinate as well as alarm. The 'uncanny' (a word that itself means beyond the system of known constructs) often exerts a strange appeal, and we derive a kind of thrill from trespassing over the edge of the safe world. Such voluntary excursions are usually harmless and usually uninformative. When, however, we wander too far, or when we are involuntarily plunged into this realm through sudden intolerable stress, whether physical or mental, then our conceptual system breaks down, and we can be overwhelmed by the unknown and threatened by madness. Or, as we shall see in Chapter 8, we may be led into the unconceptualized realm of religious experience.

The *nagual* is a sound description of the whole realm of the unknown, and Castaneda captures well the dangers inherent in our encounter with it. Our defences against the *nagual* that we build up from childhood onwards are essential to our functioning as human beings. We cannot live with constant incursions of the totally unknown. Yet, though we need our defences, our rigid and certain worlds, to protect us against being overwhelmed by the *nagual,* it is only by admitting the unknown into our lives that we can gain power and control over our lives. If we are completely cocooned against the unknown, then the slightest shock, the slightest accidental encounter with the unknown, will unbalance and weaken us. We will be like hospital patients kept in a sterile environment for long periods, who lose all their natural immunity. If, however, we are used to the unknown, then each encounter with it can enlarge our world and deepen our inner self.

Taussig, in his accounts of Colombian shamanism described earlier (see p.164), stresses that the power of the shaman derives literally from his regular encounter with the *nagual* in its most negative forms, the forces of evil and death — forces that are ever present realities to the poor of the land among whom the shaman works:

> The subtext to this attending to the poor is the subconscious cosmic battlefield of vices and virtues in which the healer gains power through the struggle with evil. The healer's power is incumbent upon a dialectical relationship with disease and misfortune. Evil empowers, and that is why a healer by necessity attends the 'poor,' meaning the economically poor and those struck by misfortune. (1986 p.158f)

Taussig goes on to develop his theme by describing the ritual used by the shaman, in his night-long intoxication with the drug *yagé*. He shows how the ritual derives its power precisely from the extent to which it disrupts the order of the *tonal:* 'the power of the ritual itself then proceeds to do its work and play through splintering and decomposing structures and cracking open meanings.' (p.441) This idea, of the encounter with the forces of negativity giving power, is also attested to by the fact that, in the training of both American and Australian shamans, a trance experience of death and dismemberment is central to gaining power, and we have already noted (see p.130) how the inner power that some refer to as the growth of the self (or the growth of the Soul) can arise from encounters with negative forces, that appear to diminish us. Taussig talks of this way as 'Walking in the space of death.' (1986, p.449)

Only by living dangerously can we avoid the conceptualized world, the *tonal,* becoming our prison instead of our protection. Castaneda's 'impeccable warrior' is one who has learnt to maintain his personal integrity while living on the edge of the unknown. Few of us are required to be in this state as a permanent vocation, but we all need to grasp, when it occurs, the opportunity to face this nothingness, to face what seems to be a death, if we are to learn how to truly live. This may involve facing the darkness inside ourselves, the 'shadow' of which we are often quite unaware; or the suffering of others against which we close off our compassion. It may involve letting go of a whole constructed world in order to allow a greater one to grow, a process that, as Mahrer has stressed, is perceived as a death of one self, to allow the birth of another. And at the highest level, the level at which our own self can be seen as part of an infinitely greater whole, it may mean embracing our own physical death.

5. Postscript on Christianity

The excursion that we have made into the seemingly wild and unfamiliar territory of shamanism will seem to many readers to have taken us into an area completely removed from Western culture and from the thinking of Western civilization; but this is not so. In fact, the ideas surveyed in this chapter are also to be found at the heart of the religion that has dominated and formed the West: Christianity.

That word has meant many things to many people. At its beginning, Christianity was a revolutionary mystery religion. Then, in a supremely ironic stroke of political genius, it was hijacked by the Emperor Constantine and has ever afterwards been used as an instrument of government repression. In order to understand Christianity, it is essential to make the effort to penetrate behind all the later accretions that have been piled on top of it, and try to move towards the words of its founder, uncertain and ambiguous though this effort may be.

If there is one thing certain regarding what was taught by Jeshua/ Jesus of Nazareth, it is that the teaching was about a state that he called the Kingdom. Another certainty, because there are no reasons for its being invented by later disciples, is that his style of public teaching was marked by the use of short provocative stories (called parables by the writers of the gospels) and by a vigorous slapstick style of humour, with his dialogues laced with absurd images of camels going through needles and people with planks in their eyes. The records that we have of the teaching and life of Jesus in those complex and heavily edited books, the gospels, contain a core of teachings about the kingdom on which all the gospels agree and which are frequently recorded as being delivered in this characteristic punchy style. While much of the narrative is threaded with later theological reflection that may be put into the mouth of Jesus, the sayings of this core are quite likely to be authentic. It is these that speak of alternate reality.

In his teaching he used both the terms 'Kingdom of God' and 'Kingdom of Heaven.' As a religious teacher in a country under military occupation, he seems to have been at pains to ensure that the term kingdom was not interpreted simply as civil resistance, and in this way he distanced himself from those who preached such a message. Since the phrase 'Kingdom of God' could be interpreted as a militaristic theocracy, it seems probable, and is consistent with the texts, that he tended to prefer the term 'Kingdom of Heaven.'

There has been a wild misunderstanding of this term in the past, when people have interpreted it as meaning a complete withdrawal from the world, or, even worse, have regarded it as referring only to a state after death. The understanding of this term has, however, been enormously clarified in recent years by the work of Neil Douglas Klotz on the origins of the term in the Aramaic language used by Jesus. The Aramaic equivalent of 'Kingdom of heaven' is *malkutha dashmaya*. Concerning the root that appears in the second word, for heaven, he writes:

> In effect, *shmaya* says that the vibration or word by which
> one can recognize the Oneness — God's name — *is* the
> universe. This was the Aramaic concept of 'heaven.' (1990)

The kingdom of heaven is the divine aspect of the universe, not something removed from the universe. Understood in this way, the origins of the phrase confirm the interpretation suggested by the reported saying that 'the kingdom of heaven is within you'; namely, that the process of entering the kingdom, which is the whole target of his teaching, is not to be understood as swearing allegiance to something external to the universe and removed far from it, but as entering into a different aspect of the universe whose gateway can be found inside ourselves. Entering, that is, into an alternative reality.

Once one accepts that this is indeed the core of Jesus's teaching, then it becomes clear that there is a strong shamanistic element in

Christianity, even though the deepest parts of the teaching go radically beyond normal shamanism. (In saying this I am not, of course, supporting the much wilder speculations of Allegri concerning the origins of Christianity in a shamanistic drug-cult.) The process whereby Jesus receives his commission to start teaching is precisely the vision-quest used in almost all cultures where shamanism is indigenous, in which the person called to be a shaman goes out into the desert and fasts in order to receive his guiding revelation from the spirit world. The most conspicuous gifts accompanying the commission are, as in classical shamanism, the ability to heal the body/mind of those that come to him for this, and a complete freedom from the shackles of conventionalism, a freedom that outrages the establishment figures whom Jesus often encounters.

One could continue with many more parallels, but here I want only to draw attention to the theme of alternative reality that repeatedly breaks through the gospel accounts, and to one particular story in which this theme stands out with particular clarity: the event known as the Transfiguration. In this story, a selected inner circle of disciples are given a vision of Jesus in a transformed state in which apparitions of two Old Testament prophets are seen beside Jesus, and he himself appears to be luminous, his divine authority shining out. Though some may doubt the authenticity of this particular event, because of the small number of witnesses and its rather carefully theological flavour, there is another episode that is traditionally linked with it, an episode that has the robust directness that seems to be the hallmark of the words of Jesus himself. The group comes down from the mountain where the revelation took place and meets with the other disciples who have been trying to cast out a demon from a possessed boy.

Whereas one might expect Jesus to step in with words of calm compassion and healing, teaching the disciples and liberating the boy, instead he bursts out in fierce and almost incomprehensible condemnation with: 'What an unbelieving and perverse generation! How long shall I be with you? How long must I endure you?' The explanation can only be found in the coupling of this incident with the preceding transfiguration episode. Jesus is standing in the alternate reality and contrasting it with the ordinary reality in which his disciples are standing; within 'the kingdom' this particular cure can be seen as an effortless flowing of authority ('Jesus spoke sternly to him; the demon left the boy'), but the disciples, despite all that they have been told, are clinging on to ordinary reality and as a result struggling hopelessly. The parallels with don Juan's outbursts at the way Castaneda clings to ordinary reality are striking.

Yet this insight into Western religion is, as I have already hinted, only half the story. While Christianity, along with many other religions, starts from a foundation of the recognition of alternate reality, it does not end there. Unlike shamanism, the great religions move deeper to a level at which they can start to make sense of the absolute ground underlying the multiplicity of alternate realities. In the next chapter I make a very sketchy beginning at exploring this new territory.

Chapter 8
Some Religious Realities

We have covered much ground, encompassing Newtonian and quantum physics, postmodernism and shamanism. All this has undermined, in various ways, the fixed concept of reality that was the foundation of the modern scientific viewpoint. Yet it has not under mined the notion of reality itself. While Mahrer allowed for the possibility of my construction of my world as a completely self contained exercise, proceeding only from my own imagination, in practice no viable way of looking at the world proceeded to this extreme. Each approach stressed, in its own way, the idea that ultimately there is something other than me; indeed, that there is an unfathomably vast realm of the non-me which is the foundation of my limited mental sorties of understanding.

Rather than destroying the notion of reality, I have tried to reinforce it by enlarging the possibility of our knowledge. Whereas conventional Newtonian science, building within the framework of Descartes, offered only a bare geometrical knowledge of the positions of particles as being real, we have now seen how this is extended by the flexible possibilities of quantum theory, and by our new relational concept of space. At the same time, the modern criticism of the scientific process has shown that this scientific conception of reality is not the last word, but only one possible framework for understanding our world.

To complement the Newtonian picture of the world I have stressed the role of consciousness, of our direct awareness of the world, and have described how, within a quantum theoretic approach to consciousness, we can allow our experiences of qualia to be accepted as giving us real information about reality. Even though it defies any fixed description, reality remains, enriched rather than destroyed by the many attacks made upon its scientific descriptions.

Yet, at the end of it all, what is the good of a concept of reality that is so nebulous that any statement about it can be denied? Does that not rather suggest that, despite all I have just said, there is no such thing as reality at all? To answer this we have to delve yet deeper into our experience, to the levels of consciousness that leave behind the world of sensation, the world of the self, that has mainly been our concern, and start to probe what lies behind the varied forms of perception and behind our individual self. It is only at this level that there is any hope of answering these questions. The transpersonal level that this takes us to is the level that has for tens of thousands of years been the subject of the religious quest.

Though it is almost absurd to attempt to say anything on this area in a single chapter, the story would be incomplete — would even give the reverse impression to what is actually the case — if I left the religious dimension untouched. So I need to end with a very cursory sketch of some of the ways in which the ideas that have been examined here interact with religious traditions, focusing on the Western religious tradition, with which I am most familiar.

I have been arguing so far that, if we are working in terms of concepts, theories, things that can be expressed in words, then the idea of a fixed reality, independent of ourselves, that we just come along and observe, cannot be maintained. While there is a 'something' other than myself, when I come to formulate this as a mode of reality, then that reality lies in relationships with me, not independently of me, or in relationships between and within particular sections of society.

Religious thinking goes beyond this. It accepts that all this is valid as far as it goes, but insists that one cannot ignore the 'something' that is more than purely individual. Religions, or at least those that I will be considering here, hold that in this realm there is indeed an absolute reality, which is open to experience, but by necessity it cannot be expressed in concepts and language, even in the new and fluid language of quantum theory.

The word 'religion' means so many different things to different people, however, that I need to begin with some statement of how I am using the word — at the risk of trespassing where angels and the best philosophers fear to tread. The subject is by definition concerned with the most deeply held views, and my use of the word 'religion' will reflect value judgments that many would strongly deny.

1. What is religion?

The history of religion has in the past often been seen through the distorting lenses of established Christian churches or the more philosophical strands of Hinduism. More recent analyses (Baring and Cashford 1991; Goodman 1988) stress, however, the roots of religion in the changing of consciousness through, for example, religious trance, so as to reveal the 'alternate reality' that we visited in Chapter 7. These practices were accompanied by stories that reinforced a conceptual framework shaping the total world in which the society lived, including the different levels of reality.

In the course of time, and within each religious tradition, the consciousness-changing practices tend to become separated from the stories. Then the stories in turn become abstracted, changed from the living experiential reality they once expressed into theories of the world and prescriptions about behaviour. When we talk about religion, in a broad sense, we are usually thinking of the situation today when these developments have taken place. In this case it is becoming customary to use 'religion' to refer to the stories — or, more generally, to the body of stories and theory that constitutes a teaching; and to refer to the consciousness-changing practices as 'spirituality.' When, however, we talk about religion in this narrower sense we must always remember the practices that, properly considered, cannot be separated from the teaching.

Therefore I think of a religion, or a particular strand within a religion, as essentially a teaching, from which flow many types of different action and practice. There are several different sorts of teaching to be found within different religious traditions, however, and it is helpful to distinguish them from each other. These types all overlap, and most teachings can be approached under any or all of the headings. Not all are relevant to 'reality.'

Myth

This is a story told within some society to make sense of the people's experience. While it has some similarity to a scientific hypothesis (e.g.

atomism at the start of the seventeenth century), its role is usually very different: it is not intended as a statement about reality, to be probed and tested by experiment, but as a story from which to draw psychological strength in order to live in that society. It can have implications as to what that society regards as the principal realities of life. For example, the second creation myth in Genesis establishes an attitude of mind relating humanity to God and to nature, and implies the prior assumption that God is a real being.

Symbol and parable

These differ from myth in that the author is consciously aware that the literal meaning is fictitious, telling the story in order either to convey a content for which there are no adequate words, or to make the message more memorable by challenging the hearers to find their own interpretation. (Jesus excelled in a combination of the two.) This mode can make statements about reality where the symbolism becomes refined to a fixed system; but more often there is a deliberate use of multiple layers of meaning, no one of which is primary. (A good example of this is alchemy, which is always simultaneously about the transformation of the world and the self). Underlying this is the view that there are natural correspondences between different areas of experience (in particular, macrocosm-microcosm) which make multiple meanings fruitful.

One might note in passing that the tradition in science has been to avoid at all cost these multiple layers of meaning, and restrict technical language as far as possible to exactly one meaning. But in computer science there has recently arisen a systematic application of multiple layers of meaning (the system of polymorphism used in object oriented programming), which opens up very interesting possibilities for understanding religious language.

Philosophy

This can be either written or oral. In contrast to the above, the aim here is to talk about reality, though symbolism is often mixed in as well — there is a continuum between the two types. Eastern writings (examples best known to Western readers being the *Bhagavad Gita,* and Lao Tze) are closer to this mode than Western ones, where it is usually found in secondary texts only.

Mysticism

These are teachings, in the form of any one of the preceding types, which stem from the teacher's claimed direct apprehension of a higher reality. They are put forward in order to lead the hearer to a similar apprehension, either by trying to describe what is experienced or, more usually, by instructing in exercises to modify consciousness in order to have the same experience. In the latter case the writings may be highly misleading if interpreted as direct accounts of reality.

The physicist E. Bastin expressed this well in describing such writings as 'training language.' He used the example of a rowing coach, who would stress that additional speed was given to the boat by the way in which the rowers carried their oars back, out of the water, after each stroke. In terms of the dynamics of the boat this was nonsense: once the oars were out of the water the boat slowed steadily down. But by behaving as if the statement were true, a style of rowing was achieved which did indeed make the boat go faster as a whole. The statement became true at a deeper level once it was acted upon.

One final distinction must be made, before examining particular religious ideas. We have been concerned, most of the time, with theories about reality: with stories about how reality is, descriptions and concepts or reality. From a mystical viewpoint, however, the aim of religion is very often not to know about reality (in a theoretical way) but to live in a state of openness to reality. This is the intended aim of the bulk of religious activity (popular devotion, moral teaching, and so on). The aim is practical, not theoretical: they are to enable one to live realistically, without necessarily imparting any intellectual knowledge of reality.

2. Shamanistic and unitary religions

I have already described (see p.175) the similarities between shamanism and Christianity, while at the same time indicating that Christianity is able to pass further than shamanism. So it will now be
helpful, as we move towards looking at religions such as Christianity,
to examine where the difference lies.

I have used examples from the shamanistic religions at various
points in order to highlight the possibility of alternate realities; and I
have also underlined the fact that, while from one viewpoint these
may be separate realities, yet from another they may be points in a
spectrum or aspects of a single reality. The distinction is important
when approaching different religions. While the shaman thinks in
terms of a world in which there are different layers of reality, between
which the shaman moves, religions that stress an openness to reality,
as I described above, are ones in which the different aspects of reality
are seen as present together, all the time. Because of this, I call the
more organized religions 'unitary' in comparison with shamanism.

While I consider this to be a difference in practical approach between shamanistic and what I might call 'unitary' religions, many
writers regard the distinction as a fundamental difference in stance.
Burnett, for instance, in his otherwise excellent *Dawning of the Pagan
Moon,* makes a clear distinction between paganism, which is based on
the shamanistic state of consciousness, and Christianity, based on the
ordinary state of consciousness. He regards the latter as superior, very
much on the basis of the 'Boss reality' thesis of De Mille.

There are two difficulties with this approach, however. One is that
it is untrue to the Christian tradition, in which, as we have seen, there
is a major place for direct apprehension of other orders of reality. The
other difficulty is that Christianity, together with other religions, also
has a tradition of the unitary approach in which the ordinary state of
consciousness can become increasingly broadened so that it includes
within it some aspects of those levels of reality that the shaman ex-
plores more explicitly. Sometimes this broadening of consciousness is
described as happening in a single moment of enlightenment, which
does not introduce new elements into the world one lives in but

transforms it by disclosing new layers of reality within each aspect of the world. As the Zen saying puts it:

> What does one do before enlightenment?
> One chops wood and carries water.
> What does one do after enlightenment?
> One chops wood and carries water.

For others, the enlargement of consciousness is not through a moment of enlightenment, but through the gradual process of spiritual growth.

3. God and the mystical

Many people, even if they admit some historical validity to the picture I sketched above (see p.180), would object to my tendentious use of the word 'stories' in defining religion, claiming that religious thought has moved beyond mere stories, and is now a repository of (more or less certain) factual truth about ultimate reality, this ultimate reality being God. A religion, they would say, is in its essence a collection of facts about the world and about God. So I need to make it clear that this is not the sort of religion I am talking about. If religion were primarily a body of facts, then it would be subject to the same critique that has already been levelled against science in this book. Religion would be, like science, a cultural phenomenon giving insight into what some particular society chooses to call reality, but would be weaker than science. While science does reach out into the realm that lies beyond the purely cultural, into the realm of the not-me, religion, if it is thought of as just a collection of facts about God and the world, would seem to be much more culturally restricted and able to tell us even less about reality than can science. The importance of religion for us in this chapter is that it goes beyond the factual into those layers of experience that I am calling the mystical.

It might be thought that, in rejecting the idea of religion as facts about God, I am following writers such as Don Cupitt (1980) who also take as their starting point the untenability of God as the sort of objective reality about which one can collect up facts. But while we start off in the same direction, initially following the same postmodern critique that reveals the inevitable subjectivity of all these alleged facts, I soon diverge from Cupitt's approach. For his reaction to the weakness of a religion that lays claim to objective facts about God is to retreat to the purely internal, the purely subjective. Religious propositions become propositions about ideals for the development of the human spirit; prayer becomes:

> a way of opening ourselves to the requirements that we have laid upon ourselves and meditating upon the ideals and values to which we have committed ourselves. ... In several senses of the phrase we pray for ourselves and we have to answer our own prayers, for it is superstition to suppose that our prayers will be answered apart from our own efforts. (1980 p.131)

I take precisely the opposite view. Recognizing the hollowness of religion as facts about God, instead of retreating to the purely subjective and human, I want to go behind the factual façade of religion to discover its mystical heart, the wisdom that sees beyond the varieties of apparent reality. The issue of prayer is a touchstone of this difference in attitude. For the primary meaning of prayer is, in traditional Christian language, the lifting up of the soul to God; only secondarily is it concerned with asking for things, or with words at all. At the heart of the Western tradition, prayer is precisely the opposite of Cupitt's account: it is going beyond the self, the ego, so as to leave behind everything that we have laid upon ourselves, or had laid upon us by others; going beyond the words and concepts that are conditioned by our particular cultural theories; and ultimately going beyond particular modes of consciousness, however valuable each may be in steering and guiding the path, to what lies beyond all concepts and all questions.

Those who regard religion as simply 'stories/teaching about God,' will be surprised that until here I have mentioned the word 'God' only once, and that in connection with myth — particularly in view of my allegiance to Christianity. So I need to comment on this, in order to make clearer what I am presupposing when I talk about religion. A modern Sufi teacher, when asked why he had not so far mentioned God or Love, replied that these were words that he did not use lightly. The first, in particular, means so many things to people of different traditions that all too often the word confuses and obscures what one is trying to say. Castaneda places the term within the *tonal,* not the *nagual,* and that is correct for the way in which most people use the word. It is in this case a particular human concept, a concept that is often not very useful. Also, it is crucial to remember that religion need not be about God: Buddhism does not use the word, while being at the same time a deep and universal religion that takes full account of the same range of experience as the religions that do talk about God. Buddhism approaches the same inexpressible areas of experience from the side of not-God as other religions (theistic religions) do from the side of God. (The idea of the Buddha-nature, for example, can come close to the Hindu idea of Atman, an aspect of God.) As I shall illustrate further later, in this attempt to go beyond the boundaries of language, several writers from theistic religions have dropped the word 'God,' some Sufi writers systematically using the word Reality *(al Haq)* as a better expression of their meaning.

I follow these writers in taking the view that religion is about reality, and that when a religious writer implies or asserts that reality is God, then we have to work hard to understand exactly what that writer means in practice, before we can understand anything of what is being said.

I have already used the word 'mystical,' which is often radically misunderstood. It has popular connotations of wizards in cloaks muttering incomprehensible alchemical formulae; and even with those who know about the religious use of the word, it often conveys the idea of some vague and shadowy apprehension, attained only after arduous training to reach a peculiarly distorted state of consciousness. What I have in mind is completely different. I am referring to any systematic exploration of our faculty of experiencing. Since our individual world is what we experience (whatever 'The World' may be in an absolute sense), an exploration of our experiencing is as vital to an understanding of our world as is an exploration of scientific ideas. This is why the study of consciousness has played such an important part in our investigation of reality, and it is why we are now examining what I call mysticism.

Because I focus on this aspect of religion, it might seem as though this was all there is to religion. In fact, in all religions, mystical experience is inseparable from morality and action. Buddhism couples right thought and right action; Hinduism — for instance in the *Bhagavad Gita* — teaches that right action, Karma Yoga, takes an essential place in the religious path; Christianity is insistent that there is no way in which one can separate love from works of love, and so on. In traditional religious language, the link between right action and experience is expressed through the idea of purification. One does not experience ultimate truth by doing certain things, but by being a certain person, by becoming open to the world. 'Purification' is the growth towards this openness, and it is a growth which is fostered by and which produces right action. In this way, right action is the prerequisite for any valid knowledge of the world. I am largely presuming this when I talk about mysticism.

To return, then, to the distinction between the popular view of mysticism as some kind of fuzzy wizardry, and religious mysticism, which is based on a refinement and heightening of the ability to experience: the common core of mystical experience is an apprehension of an order that appears *more real* (more vivid, believable, certain, absolute) than our ordinarily perceived world. This 'order' of greater reality need not be something distinct from the material world; most often it is an experience of enhanced reality in and through the same material world as we see and feel with our senses. The experience of this does indeed come to those who have undergone training; but it can come to those with little prior religious experience, and (as the work of Alister Hardy has shown) it does come to a large fraction — perhaps the majority — of the population at least a few times in their lives. The experience carries with it a conviction of solidity and reliability. It is unfortunate that all too often those who experience this are unable to use the insight or to draw any conclusions from it, because they are afraid of being laughed at.

While, however, the occurrence of mystical insight is widespread, and is open to all, in itself it need not lead to anything that is solid enough to provide the sort of test of alternative realities that I am claiming for it. Indeed, to think of mysticism as only the sort of experience that might appear spontaneously without preparation is probably as misleading as it is to think of it as wizardry. Cupitt seems to be doing this when he cites, as a typical experience of mysticism, an occasion 'triggered by the sight in blazing sunshine of a vast wall of teeming azure blossom.' He rightly argues that a one-off experience such as this is likely to be occasioned by prior theoretical (and hence cultural) considerations, and that little can be derived from it as a result.

Little, that is, if it remains in isolation. But for many people an initial experience such as this, if grasped and followed, can be a key opening the way to a journey that leads much deeper. It is this journey that I am thinking of when I speak of mysticism. The journey does not primarily consist of the experience of alternative realities (though that is part of it); but at its crucial times it is an absence of experience, called *the night of the senses,* and an absence of understanding, called *the night of the spirit,* that make up a process of discovery that lies below all experience and understanding. As Thomas Keating puts it, in this language of conventional Christian theology:

> It presupposes the gradual purification of the sense faculties
> in the night of the sense and of the spiritual faculties in the
> night of the spirit. Thus the essence of the contemplative
> path is not to be identified with psychological experiences
> of God, though these may occasionally occur.
>   ... The deeper our prayer actually is, the more it
> habitually drops out of our ordinary awareness. (1994,
> pp.45, 95)

It is this underpinning nature of mysticism that makes the mystical experience one of a realm that is more real than normal consciousness. In itself, an experience of the supra-real may not help us. We are faced with the same problems as Castaneda, with his separate reality. Do we follow De Mille's approach and designate one of the two as the 'Boss reality'? If so, which one, given that it is the normal experience that seems the most illusory?

The solution traditionally adopted is to regard the mystical experience as an insight into the Boss reality, and so to regard the reality revealed in our normal consciousness as a subsidiary reality. This then leads us to the universal image of *reality through the looking-glass.* It is an image that, like Lewis Carroll's story, opens up more twists and turns the more we look at it. The mirrors of the ancient world were polished pieces of metal that threw back a blurred and distorted image very different from the real thing. So what we perceive as ordinary reality is distorted images of the single presence that is the ultimate reality. Becoming aware of the ultimate unity of these images is like seeing that the mirror itself is a single entity. But we can also think of each object as itself a mirror of the ultimate reality. This language is used by Ibn al-'Arabi, for example, when he writes that: 'The natural order may thus be regarded [at once] as [many] forms reflected in a single mirror or as a single form reflected in many mirrors ...' (1980 p.87), and the mirror analogy is echoed in Christian writers such as Hildegard of Bingen (in particular, in her *Columba Aspexit).* As we shall see below, however, the mirror image is strongly modified in these later writers, and Ibn al-'Arabi is not here in fact saying that one order of reality is higher than another. Indeed, his concept of reality is that of a synthesis that brings together all the smaller, conflicting realities, overcoming the duality that sets them against each other. Such a synthesis is unimaginable intellectually, but can be encompassed through mystical love. For him, this synthesis of all opposites is the highest aspect of God. In particular, the 'Boss reality' of De Mille, as one reality against other lesser realities, is firmly rejected.

4. The supra-real and Platonism

Even more famous, and unambiguously fixed on the different levels of the two realities, is Plato's analogy of the perceived world as a shadow of the supra-real higher world:

And now, I said, let me show you in a figure how far our
nature is enlightened or unenlightened: Behold! human
beings housed in an underground cave, which has a long

entrance open towards the light and as wide as the interior of the cave; here they have been from their childhood, and have their legs and necks chained, so that they cannot move and can only see before them, being prevented by their chains from turning round their heads. Above and behind them a fire is blazing at a distance, and between the fire and the prisoners there is a raised way; and you will see, if you look, a low wall built along the way, like the screen which marionette players have in front of them, over which they show their puppets.

 I see.

 And do you see, I said, men passing along the wall carrying all sorts of vessels, and statues and figures of animals made of wood and stone and various materials, which appear over the wall? ...

 You have shown me a strange image, and they are strange prisoners.

 Like ourselves, I replied; for in the first place do you think they have seen anything of themselves, and of one another, except the shadows which the fire throws on the opposite wall of the cave? ... And of the objects which are being carried in like manner they would see only the shadows? ... And if they were able to converse with one another, would they not suppose that the things they saw were the real things? (1970 p.296)

The quotation, and the dialogue that follows it, show that accompanying this allegory is the value judgment that not only are we mistaken in what we call real, but that we are somehow forcibly constrained into a position that is both mistaken and pitiably wretched.

 In developing this image, Plato referred to the supra-real objects of which we perceive only the shadows as ideas. As with the word 'mysticism,' the word 'ideas' conjures up for us something shadowy and less certain than a perceived fact; but for Plato it was just the opposite. He held that the ideas were the only genuinely real things, and that after death and before birth our souls (the only genuine and authentic part of ourselves) lived in heaven in contact with the supreme reality of ideas. Only occasionally did our souls come down to the captivity of earth where reality was hidden, and where we had to make do with shadows. Plotinus, the founder of neo-platonism, elaborated this to a system with the levels of One (above being), Intellect (the realm of pure ideas), Soul (separated being), and matter (non-being). Humanity as embodied soul lives poised between Intellect and matter, with the choice of ascending to Intellect or descending to matter. Later writers assimilated the scheme to different theological positions, in which the One was identified with God as absolute, and the Intellect with the mind of God.

The idea of orders of reality 'proceeding' from the One was expressed by Christian writers influenced by neo-platonism through the idea of the *logos* (word) which was separate from the higher level but an expression of it. This way of thinking keyed in with the biblical terminology of John the Evangelist, who refers to the Cosmic Christ as *Logos,* and as God. Historically this was in part a misunderstanding, since John used the word *logos* in a very different sense, linked to the Hebrew *dhabar* and to the Old Testament theology of creation. Later writers, however, preferred to interpret it in neo-platonic terms. If the Cosmic Christ was an example of *Logos* at the highest level, at lower levels all beings were thought to contain *logoi spermatikoi,* seed-words, constituting their essential being. The *logos* is different from God but not separate from God. Hearing the word as an expression of God is a recognition that the word has its reality only as part of God; grasping this reality is thus a means of returning to God as Source of the word.

5. Theologies of affirmation and denial

Any discussion of religious teaching must tackle the problem of whether or not words and concepts can say anything about reality. This takes us to the heart of all our previous discussions. It faces up to a fundamental paradox that underlies them all: if I grasp something as fully real, then I understand it and can describe it in my own terms; but if I conceptualize and describe something in language then I am replacing the flesh-and-blood reality by a less real verbal description.

We have had many examples of this: the distinction between verbal description of our experience, which merely points to particular aspects of experience without conveying what it is really like, and the qualia which are the actual experiences themselves (and which, as we saw in Chapter 6, have a basis in quantum theory; see pp.147,) the distinction between Castaneda's *tonal,* the world of concepts, and the *nagual,* the world prior to concepts. For something to be registered by us as real, we must enter into a relationship with it; but if the relationship is one of assimilating the thing to our self, of tying it down to our terms, then its individual reality is destroyed, and is replaced by a joint reality, constructed through the interaction. If we are looking for reality unconditioned by ourselves, then we cannot grasp such a reality. For religious thought, the question is, can we be grasped by it?

Not everyone would share this sceptical view of our ability to grasp reality, however. Plato held that it was possible to escape from the imprisonment that held one looking at the shadows on the cave wall. Philosophical questioning could lay bare the pure ideas that constituted reality, and so philosophy could free us from the illusions of sense-perception. This was a dominant strand of philosophy for many centuries. Where it was incorporated into theology, the result was sometimes termed 'cataphatic theology,' a theology of affirming, through philosophical examination, what reality was like. For Platonism, and for some branches of the so-called gnosticism that flowed from it, the intellect was our primary tool in reaching an apprehension of reality. (I return to the role of the intellect later.)

This positive view of reality started to be eroded by Descartes, who, as we have seen, pointed out the extent to which our senses distort what we perceive; but he held that there were at least some ideas that gave us a handle on reality, even though most of our concepts arise from subjective mental processes. Later, even this was questioned, leaving us with few alternatives: we can become hard-nosed scientists, firmly insisting that reality is what science says it is, and that is all; we can expect less of 'reality' than in the past, and try to work to a limited concept of reality that places it in relationships; we can adopt a nihilistic position of accepting that nothing can be said, so that 'anything goes'; or we can adopt the religious position, that concepts are indeed powerless to get to reality, but that there are none the less other ways of breaking through the barrier of concepts and making contact with reality. Thus the dominant theme of mystical experience is that while concepts can point us in the right direction, the journey must be done without concepts, because no human concept can encompass absolute reality.

Traditionally, the choice has been presented as one between two different ways of using the mind: either intellectually (focusing on concepts) or mystically (focusing on experience). There is, however, a third alternative, involving the body as well as the mind. For us today, this is a vital area, because the quantum world that we examined in Chapter 6 (see p.144) has no distinction between body and mind: consciousness is just the interior part of a state of the body — a state that is pre-eminently in the brain but which could be presumed to interact fully with the rest of the body as well, since we are evolved as an integrated organism. The use of the body is well established in many religious practices, including those linked to mysticism, but it has always been handled with great caution, and has very often been kept secret because it is more open to misunderstanding and abuse than any other aspect of religion. Fortunately we are now entering a climate where the role of the body in our total life can be discussed freely, and this can start to take its proper place in religion. Because the subject is so vast, however, I shall not carry it any further here, restricting myself from now on to the conventional distinctions between different ways of using the mind. The bodily dimension, like the moral dimension, will be implicit rather than explicit.

While cataphatic theology takes the path of the intellect, the mystical route emphasizes what concepts *cannot* say, and so it is called 'apophatic,' denying, theology. Stripping oneself of concepts, which necessarily tie one down to what is less than real, involves continually saying what God is not. I want to examine some examples of this next, examples that illustrate the way in which religious thought can accept the sceptical position about reality and the inability of concepts to grasp it, which is suggested by our discussion so far, while at the same time positively asserting the possibility of being grasped by that reality.

6. Meister Eckhart

Johannes Eckhart, generally known as Meister Eckhart, was born around 1260 in Germany. While he wrote many theological treatises in Latin, it is in his German sermons to the nuns in local convents that his message is most clear, giving us a picture of the cosmos that flows from his experience of union with the ground of all being. He taught that this ground (like the One of the neo-platonists) is the ultimate source from which we come and to which we long to return.

Just as don Juan places God in the *tonal,* along with all the concepts and other junk of society, so also Eckhart is bold enough to declare that this ground of being so transcends concepts that one cannot speak of God there:

> When I stood still in my first cause, I had no God. I was cause of myself .... But when by a free will I went out and received my created being, then I had a God. Indeed, before there were creatures God was not yet God, but he was what he was.... And were it that a fly possessed reason and could intelligently seek the eternal abyss of divine being out of which it has come, then we would say that God, with all he is as God, would still be incapable of fulfilling and satisfying this fly. Therefore we beg God to rid us of God ... *(Blessed are the poor,* in Schürmann 1987, p.226)

How is one to find this ultimate reality, so deep that it even transcends what we call 'God'? Eckhart is insistent that the only way is to leave concepts behind. In his German sermon on the Bible text: 'Paul rose from the ground and with open eyes he saw nothing,' he uses the words of the text as an excuse for expounding the way of letting go of concepts so as to come to where Paul was, in his conversion on the Damascus road, when he saw God:

> I cannot see what is One. He saw nothing, that is to say, God. God is a nothingness, and yet God is a something. What God is, he is totally .... If you visualize anything or if

anything enters your mind, that is not God; indeed he is
neither this nor that. Whoever says that God is here or
there, do not trust him. The light that is God shines in the
darkness. God is a true light. To see it, one must be blind
and one must divest God of everything that there is.
(Eckhart, *Saul rose from the ground,* in Schürmann 1987)

In the same sermon he explains how thoughts can take one some way,
but a quite higher faculty, which does not involve discursive thinking,
is needed to penetrate reality:

A master says that in this light [from heaven] all powers of
the soul surpass themselves and are elevated: the exterior
sense, by which we see and hear, as well as the inner ones
which we call thoughts. How vast and unfathomable these
[thoughts] are is a marvel: I can indeed think as easily of
things beyond the sea as of what is close to me. Above the
thoughts goes the intellect in its pursuit. It goes about and
seeks; it is on the watch, here and there, gathering and
losing. But above this intellect that seeks, there is another
intellect which is not a seeker. It stands in its pure and
simple being which is inundated by that light. And I say
that in this light all the powers of the soul are super-
elevated. The senses soar up into the thoughts.

Eckhart also uses the *logos* principle that all beings can be seen as
having their ultimate reality in God. To return to the fly of our first
quotation, once the fly, and all other creatures, came out of the ground
of being, then 'God' appeared in the distinction between God the creator and God in the fly: 'But when creatures came to be and received
their created being, then God was no longer God in himself, rather he
was God in the creatures.' So we and every creature, however
insignificant, has the spark of God within: 'If we take a fly, in God,
it is more noble in God than the highest angel is in itself. This is how
all things are equal in God and are God himself' *(Qui audit me).*
   Although much of this is reminiscent of Platonism, he reverses
Platonism in his understanding of the relation between the sensory

world and God. For Plotinus the aim of meditation is to leave the sensory (non-being) behind; for Eckhart the aim is to perceive the inner *logos*-reality actually within the inner reality of the ordinary world. We do this by letting go of discursive thoughts so as to release first the 'intellect that seeks' and then the higher powers of our mind that 'stands in its pure and simple being.' Then our sensory perception of this world is drawn up into the higher intellect. To describe this process he frequently uses, as we have seen, homely images of our fellow animals, to emphasize how what he is describing is not some exotic distortion and rejection of nature, but the delightful fulfilling of what we really are:

> The seed of God is in us. Now, the seed of a pear tree
> grows into a pear tree; and a hazel seed grows into a hazel-
> tree; a seed of God grows into God. (Fox, p.28)

The message of Eckhart is, first, that there is a reality so deep that it transcends all concepts; and second, that this reality is accessible to us all if we are prepared to let go of our very selves and sink into the seeming nothingness, which is yet something, of this ultimate reality.

7. Jalal ad-Din Rumi

I have already mentioned the Sufi tradition in talking about the image of the mirror used by al-'Arabi. The origins of Sufism (first named around 750 AD) are disputed: Hadland Davis (1932) regards it as a neo-platonic sect whose members:

> with infinite licence ... quote [from the Koran], and still
> more ingeniously add their own explanations when
> necessary. No doubt there were political reasons for
> adopting this method of concealing heterodox ideas under
> the cloak of orthodoxy.

This judgment, however, misunderstands the relative roles of neo-platonic and Islamic teaching in Sufism. Though there are strong neo-platonic influences, the Sufis exhibit the same reversal of neo-platonism as did Eckhart: the aim is not removal from the world and the senses, but the perception of God within the world. This indicates that their prime motive is religious (Islamic), and that they then use the neo-platonic language as a means to a religious end. This transformation of neo-platonic language is very clear in Ibn al-‘Arabi, who freely talks about essences (that is, Plato's Ideas), but is clear that they do not exist in themselves, but only achieve existence in the cosmos. Consequently, much more convincing is Seyyed Hosein Nasr's view (1964) that the Sufis represent a strand of Islam that was originally an integral part of the faith, but which subsequently parted company from more legalistic ideas, finding a neo-platonic language in the course of doing so. It is clear from its historical origins that the Koran is a deeply mystical work, as well as a basis for life and legislation, and so it is only to be expected that both mystical and legal aspects can be drawn from it, and that initially these aspects were united. Thus, far from distorting the Koran, as Davis claimed, the Sufis are maintaining one authentic aspect of its reading.

On this view, Sufism is the mystical part of Islam. It emphasizes the power of love in drawing the soul to God leading to an ecstatic union with God. As with Eckhart, the path involves the negation of all concepts of God. What makes Sufism particularly important for our study here, however, is the stress that they lay on one of the Names of God in Islam: *al Haq,* Reality. Indeed, were it not for the orthodox teaching that the supreme Name must be Allah, it would seem from the Sufi teaching that the ultimate nature of God is simply Reality.

Perhaps the greatest poetic genius of Sufism was Rumi (1207–73). He was a respected Sufi teacher already when he underwent a trans-formational experience as a result of meeting an individual called Shams-i Tabrisi, about whom little is known. After this, his life was devoted to a constant poetic outpouring of an expression of the love of God.

Like all Sufis, he emphasizes that the central saying of Islam — *La ilaha illa'Lla* (There is no god but God) — begins with *La* (no):

All things perish, except His Face: Since you are not in His
    Face, seek not to exist:
'All things perish' no longer applies to him who is
    annihilated in Our Face,
For he is in *but God,* he has passed beyond *no god:*
    whoever is in *but* has not been annihilated.
He said *No god,* then He said *but God: No* became *but God*
    and Oneness blossomed forth.

The final line echoes Eckhart's startling: 'I pray God to rid me of
God.'

What, then, is the path to reality, if concepts are useless and there
is nothing but denial? Both Rumi and Eckhart agree: the path is love
— the love which, as we noted earlier, draws the self out to be lost
in the object, not attempting to grasp the object to the self. The path
involves following the heart in the light of this understanding: Ibn al-
'Arabi insists that the only thing large enough to contain reality is the
heart. Rumi is uncompromising in his rejection of anything less than
following the heart. Intellect is blind compared to love (an ironic
reversal of 'love is blind'); the self-denials of the ascetic follower of
religious practices in his cell are useless; when one experiences reality,
it is like a cup of wine poured by a Saki (Persian cup-bearer), which
transports one to such ecstasy that one is like a drunken profligate
compared to those who live by the intellect:

> Yesterday intellect went out with a cane in its hand and
> entered the circle of profligates: "How long will you work
> this corruption?"
>   When our Saki poured on its head a cup of wine, it broke
> down the door of the ascetic's cell: "How much more of
> this worship?"
>   It threw away its rosary and abandoned hypocrisy: "Now
> is the time for joy! How much more of this senseless
> heartache?" (in Chittick 1983)

It is uncomfortable, but inevitable, that at the end of all the questioning about reality, we are reduced to saying: 'You just know when you

meet it.' In his deeply honest book *Real Presences,* George Steiner, encountering the denial of reality by Derrida and his followers, and after fully acknowledging the logical soundness of their position, was forced to appeal to the reader's inner sense of the 'real presence.' Many philosophers have argued against the existence of qualia, and yet the sunset is red. At the logical level, the criticism is unanswerable, that by appealing to 'you just know' you are abandoning logic and argument, and opening the gates to a pure subjectivism in which anyone can say anything. Yet 'you just know' is not being said lightly. It is indeed an abandonment of logic and argument, precisely because that is the only way to proceed at this point. It is a position that can only be justified morally at the end, not the beginning, of the quest through logic and intellect, when all the false gods have been revealed as idols, when the logic of criticism has led you through the void of Derrida's scepticism.

Then the only way forward is to follow the heart. And it is here that the poetry of Rumi carries its own stamp of authenticity. Housman's test of the poem that described reality was whether or not it made his beard stand on end if he recited the poem while shaving. Rumi's poems, for many of his readers, pass this test, leaving one physically gasping at what is revealed, raising every hair of the body in a tingle of recognition. I hope that some of my readers will also share that recognition:

Be Melting Snow

Totally conscious, and apropos of nothing, he comes to see me.
Is someone here? I ask.
The moon. The full moon is inside your house.
My friends and I go running out into the street.
I'm in here, comes a voice from the house, but we aren't listening.
We're looking up at the sky.
My pet nightingale sobs like a drunk in the garden.
Ringdoves scatter with small cries *Where. Where.*
It's midnight. The whole neighbourhood is up and out in the street
thinking, *The cat-burglar has come back.*

The actual thief is there too, saying out loud,
Yes, the cat-burglar is somewhere in this crowd.
No one pays attention.

Lo, I am with you always, means when you look for God,
God is in the look of your eyes,
in the thought of looking, nearer to you than your self,
or things that have happened to you.
There's no need to go outside.
Be melting snow.
Wash yourself of yourself.

A white flower grows in the quietness.
Let your tongue become that flower.

Jalal ad-Din Rumi

Bibliography

Alexander, H.G. (ed.) *The Leibniz-Clarke Correspondence,* Manchester University Press 1956.

Aristotle, *Physics,* trans. P.H. Wicksteed and F.M. Cornford, Heinemann, London 1929.

Baring, Anne and Cashford, Jules *The Myth of the Goddess,* Arkana, UK 1991.

Blake, William "A Vision of the Last Judgment" (1810) in *William Blake's Writings,* ed. G.E. Bentley, Oxford University Press 1978, p.1007.

Bohm, D. and Hiley, B.J. *The Undivided Universe,* Routledge, London 1993.

Burnett, D. *Dawning of the Pagan Moon,* Monarch, Eastbourne, UK 1991.

Castaneda, Carlos *Tales of Power,* Hodder & Stoughton, London 1975.

Chatwin, Bruce *The Songlines,* Viking Penguin, New York 1988.

Chittick, W.C. *The Suffi Path of Love,* State University of New York Press 1983.

Cupitt, Don *Taking Leave of God,* SCM Press, UK 1980.

Davis, H. *The Persian Mystics: Jalálu'd-dín Rúmí,* John Murray, London 1907.

De Mille, C. *Castaneda's Journey,* Capra Press, Santa Barbara 1976.

Dennett, Daniel *Consciousness Explained,* Allen Lane, London 1991.

Descartes, René *The Cambridge Companion to Descartes,* ed. J. Cottingham, Cambridge University Press 1992.

D'Espagnat, Bernard *In Search of Reality,* Springer-Verlag, New York 1983.

Eddington, Arthur S. *The Nature of the Physical World,* Cambridge University Press 1928.

Elkin, A.P. *The Australian Aborigines,* 3rd ed., Angus & Robertson, Sydney 1954.

Feyerabend, Paul *Against Method,* Verso, London 1975.

Fox, Matthew *Meditations with Meister Eckhart,* Bear & Co., Santa Fe, New Mexico 1983.

——, *The Coming of the Cosmic Christ,* Harper & Row, San Francisco 1988.

Goodman, Felicitas D. *Ecstasy, Ritual and Alternate Reality,* Indiana University Press, Bloomington 1988.

——, *Where the Spirits Ride the Winds,* Indiana University Press, Bloomington 1990.

Ho, Mae-Wan *The Rainbow and the Worm,* World Scientific, Singapore 1993.

Ibn al-'Arabi, *The Bezels of Wisdom,* trans. R.W.J. Austin, SPCK, London 1980.

Katz, R. "Education and transformation," *Harvard Educational Review,* 51, 1981, pp.57–78.

Keating, Thomas *Intimacy with God,* Crossroad, New York 1994.

Keller, E. Fox *Reflections on Gender and Science,* Yale UP, New Haven 1985.

Klotz, Neil Douglas *Prayers of the Cosmos,* Harper San Francisco 1990.

Kuhn, T. S. *The Structure of Scientific Revolutions,* University of Chicago Press 1962.

Lawson, Hilary and Appignanesi, Lisa *Dismantling Truth,* Weidenfeld and Nicholson, London 1989.

Lockwood, M. *Mind, Brain and Quantum: the Compound I,* Blackwell, Oxford 1989.

Lucretius, *De Rerum Natura,* trans. C. Bailey, Oxford University Press 1947.

Mahrer, Alvin K. *Experiencing: a humanistic theory of psychology and psychiatry,* University of Ottawa Press 1989.

Miller, Joan "Transforming activity," in *Theoria to Theory,* 8, 1974, pp.123–42.

Mindell, Arnold *The Shaman's Body,* Harper Collins, New York 1993.

Nasr, Seyyed Hosein *Three Muslim Sages,* Caravan Books, New York 1964.

Needleman, Jacob *Lost Christianity,* Element Books, Rockport MA 1993.

Nunn, C., Blott, B. and Clarke, C.J.S. "Collapse of a quantum field may affect brain function," in *Journal of Consciousness Studies,* 1, 1994, pp.127–39.

Penrose, Roger *The Emperor's New Mind,* Oxford University Press 1989.

Plato *The Republic,* trans. B.J. Jowett, Sphere Books, UK 1970.

Rorty, Richard "Science as solidarity," in Lawson and Appignanesi 1989.

Roszak, Theodore *The Voice of the Earth,* Transworld Publishing, London 1992.

Rumi, Jalalu'd-din *The Open Secret: versions of Rumi,* trans. John Moyne and Coleman Barks, Threshold Books, Putney, Vermont 1984.

Schürmann, R. *Meister Eckhart, Mystic and Philosopher,* Indiana University Press, Bloomington 1987.

Sorabji, R. *Time, Creation and the Continuum,* Duckworth, London 1983.

Spretnak, Charlene *States of Grace,* Harper San Francisco 1991.

Starhawk *Dreaming the Dark,* Beacon Press, Boston 1988.

Steiner, George *Real Presences,* Faber and Faber, London 1989.

Taussig, M.T. *Shamanism, Colonialism and the Wild Man,* University of Chicago Press 1986.

Tomlinson, Hugh "After truth: postmodernism and the rhetoric of science," in Lawson and Appignanesi 1989.

Wakefield, N. *Postmodernism: the twilight of the real,* Pluto, London 1990.

Walker, Evan "The Nature of Consciousness," *Mathematical Biosciences,* 7, 1970, pp.131–78.

——, "Quantum Mechanical Tunnelling in Synaptic and Ephaptic Transmission," *Int. J. Quantum Chem.* 11, 1977, pp.103–27.

Wildiers, N.M. *The Theologian and his Universe,* Seabury Press, New York 1982.

Zohar, Danah *The Quantum Self,* Bloomsbury, London 1991.

Index